COOPERATIVE EXTENSIONS OF THE BAYESIAN GAME

Series on Mathematical Economics and Game Theory

Series Editor: Tatsuro Ichiishi *(Hitotsubashi University)*

Editorial Advisory Board

James S. Jordan
The Penn State University

Ehud Kalai
Northwestern University

Semih Koray
Bilkent University

John O. Ledyard
California Institute of Technology

Richard P. McLean
Rutgers University

Dov Monderer
The Technion

Bezalel Peleg
The Hebrew University of Jerusalem

Stanley Reiter
Northwestern University

Dov E. Samet
Tel Aviv University

Timothy Van Zandt
INSEAD

Eyal Winter
The Hebrew University of Jerusalem

Itzhak Zilcha
Tel Aviv University

Published

Vol. 1: Theory of Regular Economies
by Ryo Nagata

Vol. 2: Theory of Conjectural Variations
by C. Figuières, A. Jean-Marie, N. Quérou & M. Tidball

Vol. 3: Cooperative Extensions of the Bayesian Game
by Tatsuro Ichiishi & Akira Yamazaki

Series on Mathematical Economics and Game Theory

Vol. 3

COOPERATIVE EXTENSIONS OF THE BAYESIAN GAME

Tatsuro Ichiishi
Hitotsubashi University, Japan

Akira Yamazaki
Hitotsubashi University, Japan

World Scientific

NEW JERSEY • LONDON • SINGAPORE • BEIJING • SHANGHAI • HONG KONG • TAIPEI • CHENNAI

Published by

World Scientific Publishing Co. Pte. Ltd.
5 Toh Tuck Link, Singapore 596224
USA office: 27 Warren Street, Suite 401-402, Hackensack, NJ 07601
UK office: 57 Shelton Street, Covent Garden, London WC2H 9HE

Library of Congress Cataloging-in-Publication Data
Ichiishi, Tatsuro.
    Cooperative extensions of the Bayesian game / by Tatsuro Ichiishi &
Akira Yamazaki.
    p. cm. -- (Series on mathematical economics and game theory ; v. 3)
Includes bibliographical references and index.
ISBN-13 978-981-256-359-0
ISBN-10 981-256-359-8
1. Game theory. 2. Bayesian statistical decision theory. I. Yamazaki, Akira, 1942- II. Title.

HB144.I27 2006
519.5'42--dc22

2006046418

British Library Cataloguing-in-Publication Data
A catalogue record for this book is available from the British Library.

Copyright © 2006 by World Scientific Publishing Co. Pte. Ltd.

All rights reserved. This book, or parts thereof, may not be reproduced in any form or by any means, electronic or mechanical, including photocopying, recording or any information storage and retrieval system now known or to be invented, without written permission from the Publisher.

For photocopying of material in this volume, please pay a copying fee through the Copyright Clearance Center, Inc., 222 Rosewood Drive, Danvers, MA 01923, USA. In this case permission to photocopy is not required from the publisher.

Printed in Singapore

*To Barbara, Yukiko, Wanda,
and the memory of LeRoy*

Contents

Preface   xi

I   BASIC INGREDIENTS

1 Introduction   3

2 Cooperative Extension of the Bayesian Game   7
    2.1   Formal Model   7
    2.2   Examples   13
    2.3   Two Approaches to Formulate Incomplete Information . . .   16

3 Measurability as Feasibility of Individual Actions   21
    3.1   General Case   21
    3.2   Private Information Case, Fully Pooled Information Case .   22

4 Bayesian Incentive Compatibility as Feasibility of Execution of Contracts   25
    4.1   Private Information Case   25
    4.2   Mediator-Based Approach   37
    4.3   Communication Plan as a Part of a Strategy   40

II   SOLUTIONS, INFORMATION REVELATION

5 Descriptive Solution Concepts   45
    5.1   *Interim* Solution Concepts   45
    5.2   *Ex Ante* Solution Concepts   52
    5.3   Other Interactive Modes   54
    5.4   Coexistence of Coalitions   54

6	**Normative Solution Concepts**	**57**
	6.1 *Interim* Efficiency Concepts	57
	6.2 Coexistence of Coalitions	62
7	**Comparisons of Several Core Concepts**	**65**
	7.1 Fine Core and *Ex Post* Core	65
	7.2 Private Measurability and Bayesian Incentive Compatibility	70
8	**Existence**	**91**
	8.1 *Interim* Solutions	91
	8.2 *Ex Ante* Solutions	93
	8.A Appendix to Chapter 8	98
9	**Approaches to Information Revelation**	**109**
	9.1 By Contract Execution	110
	9.2 By Contract Execution: A Profit-Center Game with Incomplete Information	118
	9.3 By Choosing a Contract	135
	9.4 Update of *Interim* Probabilities	138
	9.5 By Credible Transmission of Information During the Contract Negotiation	148
	9.A Appendix to Chapter 9	149

III PURE EXCHANGE ECONOMY

10	**Existence**	**167**
	10.1 *Interim* Solutions	168
	10.2 *Ex Ante* Solutions	188
11	**Large Economy**	**195**
	11.1 Large Bayesian Pure Exchange Economy	195
	11.2 *Interim* Solutions	196
	11.3 *Ex Ante* Solutions	201
12	**Core Convergence/Equivalence Theorems**	**207**
	12.1 *Interim* Solutions	207
	12.2 *Ex Ante* Solutions	212

IV ANOTHER VIEWPOINT

13 Self-Selection in Anonymous Environments   **219**
    13.1 Mechanism Design 219
    13.2 Pure Exchange Economy 223

Bibliography   **227**

Glossary   **235**

Index   **239**

Preface

Our collaboration on the research area, strategic cooperative game theory with asymmetric information, started around the spring of 2000, and progressed substantially from July 2001 through September 2002 when Ichiishi was visiting Hitotsubashi University (Yamazaki's affiliation). Through innumerable discussions during this period not only on our ongoing joint research topics but also on varieties of the literature in the area, we gradually formed and solidified our own common view about details of this theory, and at the same time developed the desire to present it systematically. The present monograph is an outgrowth of these initial efforts. We were eventually placed in an ideal environment for our collaboration as Ichiishi joined the Faculty of Hitotsubashi University in April 2005.

We have also "tried out" our results and our philosophy at: Decentralization Conference (held at Waseda University, October 6, 2001); Symposium of the Research Center for Mathematical Economics (held at Kyoto University, December 7-9, 2001); The Spring 2002 Meeting of the Japanese Economic Association (held at Otaru University of Commerce, June 15, 2002); The First Illinois Economic Theory Workshop (University of Illinois at Urbana-Champaign, April 18-20, 2003); and also at invited seminars at Kobe University, Tokyo Institute of Technology, and Kyoto University, in addition to our own seminar at Hitotsubashi University. To all the participants in our talks, we would like to express our gratitude. We would also like to thank World Scientific Co. Ltd. for publishing this monograph in this beautiful book format, in particular for their patience with our very slow writing.

Tatsuro Ichiishi and Akira Yamazaki
Kunitachi, Tokyo

Part I
BASIC INGREDIENTS

Chapter 1

Introduction

Since the 1970s, we have seen voluminous literature on the analysis of economic problems with asymmetric information. Harsanyi's (1967/68) Bayesian game and Bayesian equilibrium have served as a conceptual foundation for these analyses. While the literature provided new insights into the workings of the present-day economy which the traditional neoclassical paradigm failed to analyze, many works actually postulated a quite restrictive mode of players' interaction, that is, the principal-agent relationship (Stackelberg's leader-follower relationship), a specific instance of the noncooperative game.

Parallel to the development of Bayesian analyses of the noncooperative game, there has been development in static descriptive cooperative game theory, the theory which analyzes another interactive mode in which several players, with all their diverse (most likely conflicting) interests, come to form a coalition to make a coordinated choice of strategies, because by doing so everybody in the coalition ends up better off than he would behaving alone (noncooperatively). Aumann and Peleg's (1960) model of non-side-payment game (NTU game) and Scarf's core nonemptiness theorem for this game (see, e.g., Scarf (1973, theorem 8.3.6, p. 211)) serve as a breakthrough in the literature. It was with this model that economists could start analyzing cooperative behavior without imposing problematic conditions on utilities, such as the cardinal nature or the transferability. Scarf's theorem is a milestone in studies of the core, a central descriptive cooperative solution.

Wilson (1978) pioneered the study of cooperative behavior in the pure exchange economy with asymmetric information. Then came a revival of research in this area around 1990, and subsequent lively contributions by many authors established the cooperative game theory with asymmetric information as one of today's central and burgeoning research areas in economic theory and game theory. These works have been done within various frameworks, with diverse degrees of generality. The most general framework may be called a *cooperative extension of the Bayesian game*, on account of its rich structure: Like the Bayesian game it addresses players' strategy-choice in the presence of asymmetric information, but unlike the Bayesian game it allows for coordinated strategy-choice by several players. Like the non-side-payment game it explicitly formulates coalitional attainability, but unlike the non-side-payment game it allows for the influence of outsiders' behavior upon each coalition. This framework is indispensable, for example, in the analysis of an economy with organizations as production units (firms), in particular in the analysis of resource allocation mechanisms instituted in firms as superior alternatives to the incomplete market mechanism. Firms in the present-day free societies are places within which human-resource holders having private information coordinate their strategies, and these firms are interdependent in that the feasibility and implications of strategy-choice within a firm are influenced by the outsiders' strategy-choice.

Actually there have been mainly two alternative approaches to the Bayesian cooperative game: One strand is based on the view that, while the conventional noncooperative Bayesian analyses sometimes have assumed the presence of a mediator for the firm activities, there is no need for a mediator in the cooperative Bayesian analysis. Indeed, in reality, firms are operated without consulting with a mediator; the managers at various levels of the corporate hierarchy are not mediators but players in a coalition pursuing their own interests. While the principal-agent theory explains the institution of a mechanism as a solution to the mediator's optimization problem, cooperative Bayesian analysis explains it as an endogenously determined strategy bundle chosen by the insiders of the coalition. The other strand stands closer to the conventional noncooperative Bayesian analysis; it retains a mediator and gives him a major role in the coalition's execution of a strategy bundle.

The purpose of this book is to systematically present the cooperative game theory with asymmetric information – from Wilson (1978) to the current frontier, from general models to specific examples, and all the approaches that have been proposed to date. This theory is far from complete; on the contrary, there are many unsolved questions, and in fact up to now there has not been any definitive general work. Nevertheless, we decided to take on this task, hoping that clarification of the past accomplishments of the area and their appraisal will facilitate further development of the theory.

1: Introduction 5

The book is organized as follows: Part I provides the key ingredients, such as the basic one-shot model and several examples (chapter 2), and the two meaningful conditions that endogenous variables (strategies) are required to satisfy: the measurability requirement (chapter 3) and the Bayesian incentive compatibility requirement (chapter 4).

Part II reviews several basic issues that have been addressed with solution concepts. The emphasis here is on several descriptive solution concepts (defined in chapter 5), although normative solution concepts that formalize efficiency are also defined (chapter 6). We report works on the comparison of two descriptive solutions, the fine core and the *ex post* core, and on the comparison of implications of the two required conditions, measurability and Bayesian incentive compatibility (chapter 7). Given the general framework, we present existence theorems for descriptive solutions (chapter 8). We also present our view on how to evaluate the existence and the nonexistence results (the end of chapter 8). We close part II with analyses of information revelation processes, that is, studies of how private information turns into public information (chapter 9).

Part III reviews studies of various versions of the core, given the specific framework of the Bayesian pure exchange economy. There are fairly general or clear-cut existence theorems, taking advantage of the specific structure of the model (chapter 10). We also look into a large economy, formulated with a nonatomic space of consumers, in the context of incomplete information (chapter 11). Actually some works on a large economy were presented in chapter 7 or will be presented in chapters 12 and 13, but the emphasis in chapter 11 is on a solution which satisfies the two major requirements (chapters 3 and 4). We provide our critical review of the recent revival of the core convergence theorem within the framework of the Bayesian pure exchange economy (chapter 12).

Part IV reviews another view on coalition formation. Specifically, we review analyses of situations in which coalitional membership is anonymous (chapter 13).

Chapter 2

Cooperative Extension of the Bayesian Game

This chapter introduces the basic one-shot model of Bayesian society; it synthesizes Harsanyi's (1967/68) Bayesian game and Aumann and Peleg's (1960) non-side-payment game (NTU game). This set of exogenous data is common knowledge to the players of the game. Here, player j's private information is formulated as his type, so the probability space of states which formulates incomplete information consists of the type profiles. The chapter then introduces the endogenous variable of strategy, that is, a plan of state-contingent actions. Section 2.1 provides a general presentation, and section 2.2 provides three specific examples: a Bayesian pure exchange economy, a Bayesian coalition production economy, and the internal organization of a firm in multidivisional form with incomplete information. Some authors have adopted the general framework of arbitrary probability space to formulate incomplete information (rather than the type-profile space). Section 2.3 briefly reviews Jackson's (1991) observation of the equivalence of the type-profile approach and the general approach.

2.1 Formal Model

We present a cooperative extension of Harsanyi's (1967/1968) Bayesian game in this section. For full analyses of cooperative behavior, the required model needs to treat at least the strategy concept and coalitional attainability concept explicitly, so that it embodies both the ingredients of the Bayesian game and the ingredients of Aumann and Peleg's (1960) non-side-payment game (NTU game).

Let N be a finite set of *players*. The family of nonempty *coalitions* (nonempty subsets of N) is denoted by $\mathcal{N}$. Each player j has a *choice set* (an *action set*) C^j, a *type set* T^j, and a type-profile dependent *von Neumann-Morgenstern utility function* u^j. At a particular stage of the game player j alone knows which member of the set T^j is truly realized; in this sense the realized member t^j is called j's *private information*. The type set T^j will be assumed to be finite throughout this book.

For $S \in \mathcal{N}$, define $C^S := \prod_{j \in S} C^j$, $T^S := \prod_{j \in S} T^j$, and write $C := C^N$, $T := T^N$ for simplicity. A member of T^S is called a *type profile* of coalition S. Generic elements of sets C^S, T^S, C and T are denoted by c^S $(:= (c^j)_{j \in S})$, t^S, c and t, respectively.

The utility function $u^j : C \times T \to \mathbf{R}$ associates j's utility level $u^j(c, t)$ to each choice bundle c and type profile t.

The *ex ante* period is defined as the period in which each player does not have his private information, but has an *ex ante* probability on the type profiles T, subjective or objective. An *interim* period (or rather, an *in mediis* period) is defined as a period in which each player already has his private information but does not know the true type profile. Sometimes the *interim* period is more specifically defined as the period in which each player has only his private information and nothing more; in the following, whenever confusion is likely, we will clarify the meaning of the term "*interim*". The *ex post* period is defined as the period in which everybody knows the true type profile.

Given private information t^j, player j holds an *interim* probability $\pi^j(\cdot \mid t^j)$ on the others' type profiles $T^{N \setminus \{j\}}$; the probability could be objective or subjective. Sometimes we assume that it is derived from an *ex ante* probability π^j on T by the Bayes rule,

$$\pi^j(t^{N \setminus \{j\}} \mid t^j) = \frac{\pi^j(t^{N \setminus \{j\}}, t^j)}{\pi^j(T^{N \setminus \{j\}} \times \{t^j\})}.$$

Some works have an *ex ante* probability as a given datum, while others have the more general approach in which *interim* probabilities are given data. An *interim* probability $\pi^j(\cdot \mid t^j)$ is synonymously called a conditional probability given t^j.

DEFINITION 2.1.1 (Harsanyi, 1967/1968) A *Bayesian game* is a list of specified data,

$$\{C^j,\ T^j,\ u^j,\ \{\pi^j(\cdot \mid t^j)\}_{t^j \in T^j}\}_{j \in N}.$$

The set T^S gives rise to the partition of T: $\{\{t^S\} \times T^{N \setminus S} \mid t^S \in T^S\}$. Denote by $\mathcal{T}^S$ the algebra on T generated by this partition, and set $\mathcal{T}^j := \mathcal{T}^{\{j\}}$. We call $\mathcal{T}^j$ player j's *private information structure*.

2: Cooperative Extension of the Bayesian Game

Players may have more than the private information in the *interim* period. One of the factors which contribute to this information-revelation is the probabilities $\pi^j(\cdot \mid t^j)$, $t^j \in T^j$ (other possible factors will be discussed in chapters 3 and 9): When type $\bar{t}^j$ is realized, player j knows that the others' types cannot be $t^{N\setminus\{j\}}$ if $\pi^j(t^{N\setminus\{j\}} \mid \bar{t}^j) = 0$. Denote by $\operatorname{supp} \pi^j(\cdot \mid t^j)$ the support of $\pi^j(\cdot \mid t^j)$,

$$\operatorname{supp} \pi^j(\cdot \mid t^j) := \{ t^{N\setminus\{j\}} \in T^{N\setminus\{j\}} \mid \pi^j(t^{N\setminus\{j\}} \mid t^j) > 0 \}.$$

Then the set of all type profiles with a positive probability is given by

$$T(\pi^j) := \bigcup_{t^j \in T^j} \{t^j\} \times \operatorname{supp} \pi^j(\cdot \mid t^j).$$

Player j is endowed with the information structure,

$$\mathcal{T}^j(\pi^j) := \mathcal{T}^j \bigvee \{\emptyset, T(\pi^j), T \setminus T(\pi^j), T\}.$$

Needless to say, the refinement of information structure, from $\mathcal{T}^j$ to $\mathcal{T}^j(\pi^j)$, has a different meaning, depending on whether the conditional probabilities, $\pi^j(\cdot \mid t^j)$, $t^j \in T^j$, are objective or subjective.

If $\pi^j(\cdot \mid t^j) \gg 0$ for all t^j, the probabilities do not reveal information, so $\mathcal{T}^j(\pi^j) = \mathcal{T}^j$. If this is the case for all players, the model exhibits the strongest form of informational asymmetry,

$$\mathcal{T}^i(\pi^i) \bigwedge \mathcal{T}^j(\pi^j) = \{\emptyset, T\}, \text{ if } i \neq j.$$

For analysis of cooperative behavior, one first needs to distinguish between feasible and infeasible coalitional choices. When the players' true type profile is given as t, each coalition S has a set of *feasible* joint choices, defined as subset $\mathbf{C}_0^S(t)$ of C^S. Write for simplicity, $\mathbf{C}_0^j(t) := \mathbf{C}_0^{\{j\}}(t)$. Thus the *feasible-choice correspondence* $\mathbf{C}_0^S : T \to C^S$ is defined for every coalition S. Notice the possibilities that $\mathbf{C}_0^S(t) \neq \prod_{j \in S} \mathbf{C}_0^j(t)$, and that $\mathbf{C}_0^S(t) \neq \mathbf{C}_0^S(t')$ if $t \neq t'$.

Complete information is defined as the situation in which there is no informational problem, i.e., each set T^j is a singleton. In this case we suppress the notation t from $u^j(c,t)$ and $\mathbf{C}_0^S(t)$, and simply write $u^j(c)$ and C_0^S. If, furthermore, player j's utility function depends only upon his choice (i.e., $u^j(c) = u^j(c^j)$), then the set

$$\tilde{V}(S) := \{(u_j)_{j \in S} \in \mathbf{R}^S \mid \exists\, c^S \in C_0^S : \forall\, j \in S : u_j \leq u^j(c^j)\}$$

is the set of all utility allocations attainable in coalition S. By coordinated choice, the members of coalition S can realize any utility allocation in

$\tilde{V}(S)$. For different coalitions S and T, sets $\tilde{V}(S)$ and $\tilde{V}(T)$ lie in different Euclidean spaces, so it is analytically convenient to introduce the cylinders in the same space $\mathbf{R}^N$ based on $\tilde{V}(S)$ and $\tilde{V}(T)$. Define, therefore,

$$V(S) := \{u \in \mathbf{R}^N \mid (u_j)_{j \in S} \in \tilde{V}(S)\}.$$

DEFINITION 2.1.2 (Aumann and Peleg, 1960) A *non-side-payment game* (*NTU game*, or *non-transferable utility game*) is a cylinder-valued correspondence from $\mathcal{N}$ to $\mathbf{R}^N$, i.e., a correspondence $V : \mathcal{N} \to \mathbf{R}^N$ such that

$$[u, v \in \mathbf{R}^N, \ \forall \, j \in S : u_j = v_j] \implies [u \in V(S) \text{ iff } v \in V(S)].$$

A strategy of a player specifies his choice contingent upon a type profile. It is synonymously called a *plan*. Formally, a *strategy* of player j is a function $x^j : T \to C^j$, or its restriction to an appropriate domain (subset of T). A choice $c^j \in C^j$ may be identified with the constant strategy $t \mapsto c^j$.

When coalition S forms, the members get together, talk to each other, and coordinate their strategy choice. They agree on their actions for each possible type profile (contingency). Denote by $T(S)$ the set of all contingencies for which the members plan their actions; the set $T(S)$ is the domain of their strategies. Precise choice of this domain $T(S)$ should be determined by the specific context of a game; there cannot be a general theory. Here, we only impose the mildest requirement on $T(S)$:

$$\bigcup_{j \in S} T(\pi^j) \subset T(S) \subset T.$$

Definition of $T(S)$ as a proper subset of T becomes particularly important, when a general state space is arbitrarily given to the model, and this state space is re-formulated as a type-profile space so that the present theory applies; see section 2.3. In this case, all players know that some type profiles will never be realized, so they do not plan their actions contingent on such profiles. (For now until the end of section 2.2, however, the reader may assume for simplicity that $T(S) = T$ for all $S \in \mathcal{N}$.)

Denote by $X^j(S)$ the set of all logically conceivable strategies of player j as a member of coalition S,

$$X^j(S) := \{x^j : T(S) \to C^j\}.$$

For coalition S, define $X^S := \prod_{j \in S} X^j(S)$; it is the set of all logically conceivable coordinated strategies. (So, $X^{\{j\}} = X^j(\{j\})$.) Write $X := X^N$ for simplicity. The model under construction explains which strategy bundle $x^S := (x^j)_{j \in S}$ is agreed upon by the members of coalition S.

We next introduce the feasible-strategy concept. Suppose the grand coalition is entertaining possible strategy bundle $\bar{x} \in X$. If coalition S is to deviate from N in this situation, the members of S have to know which strategy bundles x^S are feasible for them. The *a priori* given set $F^S(\bar{x})$ ($\subset X^S$) describes precisely these feasible strategies. Feasibility is thus formulated by a feasible-strategy correspondence $F^S : X \to X^S$, which associates to each strategy bundle $\bar{x} \in X$ (which may not be feasible) the set $F^S(\bar{x})$ of all feasible strategy bundles available to coalition S. A feasible strategy bundle as a function $x^S : T(S) \to C^S$ cannot depend on the outsiders' types $t^{N \setminus S}$, since the members of S do not know them. So the strategy bundle x^S depends only upon their own types t^S, that is, it is $\mathcal{T}^S$-measurable. Here, the correspondence $F^S : X \to X^S$ is interpreted as the basic feasibility determined by resource constraints and the characteristics of individual players. The $\mathcal{T}^S$-measurability should not be interpreted as an informational restriction; the latter will be discussed later. Indeed, we will see in chapters 3 and 9 stronger measurability requirements reflecting information available to each member.

One might simply wish to use the constant correspondence F_0^S defined by

$$F_0^S(\bar{x}) := \{x^S \in X^S \mid x^S \text{ is a } \mathcal{T}^S\text{-measurable selection of } \mathbf{C}_0^S|_{T(S)}\}$$

as the feasible-strategy correspondence. We take another approach, however, in which the correspondence F^S is arbitrarily given, provided

$$F^S(\bar{x}) \subset \{x^S \in X^S \mid x^S \text{ is a } \mathcal{T}^S\text{-measurable selection of } \mathbf{C}_0^S|_{T(S)}\},$$

for all $\bar{x} \in X$. This allows us to take into account the possibility that coalition S's feasibility is influenced by the outsiders' strategy choice.[1]

DEFINITION 2.1.3 (Ichiishi and Idzik, 1996) A *Bayesian society* is a list of specified data

$$\mathcal{S} := \left(\{C^j, T^j, u^j, \{\pi^j(\cdot \mid t^j)\}_{t^j \in T^j}\}_{j \in N}, \{\mathbf{C}_0^S, T(S), F^S\}_{S \in \mathcal{N}} \right)$$

of: (i) a finite set of players N; (ii) a choice set C^j, a finite set of types T^j, a von Neumann-Morgenstern utility function $u^j : C \times T \to \mathbf{R}$, and a conditional probability $\pi^j(\cdot \mid t^j)$ on the others' type profiles $T^{N \setminus \{j\}}$ given private information $t^j \in T^j$, objective or subjective, for each player j; (iii) a feasible-choice correspondence $\mathbf{C}_0^S : T \to C^S$, a domain of strategies $T(S)$

---

[1] There is redundancy in definitions of $\mathbf{C}_0^S$ and F^S; once we have the feasible-strategy correspondence concept, we can consistently define a feasible-choice correspondence concept.

satisfying $\bigcup_{j \in S} T(\pi^j) \subset T(S) \subset T$, and a feasible-strategy correspondence $F^S : X \to X^S$ for each coalition S, such that each element of $F^S(\bar{x})$ is a $\mathcal{T}^S$-measurable selection of $\mathbf{C}_0^S|_{T(S)}$.

We postulate throughout this book that data $\mathcal{S}$ is common knowledge to the players N.

We clarify timings of strategy-design and of action. The members of a coalition may endogenously determine their strategy bundle either in the *interim* period or in the *ex ante* period, depending upon the nature of the particular issue addressed. The above definition 2.1.3 is appropriate for strategy design in the *interim* period. Majority of the researches done to date are, however, on *ex ante* determination of a strategy bundle, so we consider the situation in which each player j has an *ex ante* probability π^j on T, objective or subjective. In this case, a Bayesian society reduces to a list of specified data, $\left(\{C^j, \mathcal{T}^j, u^j, \pi^j\}_{j \in N}, \{\mathbf{C}_0^S, T(S), F^S\}_{S \in \mathcal{N}} \right)$. Definitions of $T(\pi^j)$ and $T(S)$ in this *ex ante* probability case are analogous to the *interim* probability case; in particular, $T(\pi^j)$ is simply the support of the probability π^j.

In accordance with the agreed upon strategy bundle, x^S, each member j then execute his strategy, that is, takes the action $x^j(t^S)$ upon realization of type profile t. Thus, actions take place in the *interim* period, regardless whether the strategy bundle was determined in the *interim* period or in the *ex ante* period.

Wilson (1978) pioneered study of the core of a pure exchange economy with incomplete information. He introduced and carefully discussed several core concepts that would allow for the phenomenon of adverse selection. We recall that adverse selection undermines opportunities for insurance in a market, as Akerlof (1970) demonstrated in his model of the market for lemons. Wilson emphasized the role of revelation of private information. When player j is endowed only with his private information structure, he can distinguish two states (i.e., two type profiles), t and t', iff there exists an event $E \in \mathcal{T}^j$ such that $t \in E$ and $t' \notin E$. Likewise, when coalition S is formed and each member somehow fully reveals his information to his colleagues, any member of S can distinguish two states using the pooled information structure $\mathcal{T}^S$. In general, each member j receives only partial information from his colleagues, so the information structure he can use is an algebra $\mathcal{A}^j$ which is finer than his private information structure but is coarser than the fully pooled information structure.

DEFINITION 2.1.4 (Wilson, 1978) A *communication system* for coalition S is an #S-tuple of algebras $\{\mathcal{A}^j\}_{j \in S}$ on T such that

$$\forall \, j \in S : \mathcal{T}^j \subset \mathcal{A}^j \subset \mathcal{T}^S.$$

It is called *null*, if $\mathcal{A}^j = \mathcal{T}^j$ for every $j \in S$. It is called *full*, if $\mathcal{A}^j = \mathcal{T}^S$ for every $j \in S$.

2.2 Examples

We present several economic examples of the Bayesian society (definition 2.1.3).

EXAMPLE 2.2.1 A *Bayesian pure exchange economy*

$$\mathcal{E}_{pe} := \left\{ X^j, T^j, u^j, e^j, \{\pi^j(\cdot \mid t^j)\}_{t^j \in T^j} \right\}_{j \in N}$$

is an economy with l commodities, where N is a finite set of consumers, and for each consumer j, X^j ($\subset \mathbf{R}_+^l$) is his consumption set,[2] T^j is his finite type set, $u^j : X^j \times T \to \mathbf{R}$ is his type-profile dependent von Neumann-Morgenstern utility function, $e^j : T^j \to \mathbf{R}_+^l$ is his initial endowment vector, which depends only upon t^j, and $\pi^j(\cdot \mid t^j)$ is a conditional probability on $T^{N \setminus \{j\}}$ given t^j, objective or subjective.

The associated Bayesian society

$$\left(\left\{ C^j, T^j, u^j, \{\pi^j(\cdot \mid t^j)\}_{t^j \in T^j} \right\}_{j \in N},\ \{\mathbf{C}_0^S, T(S), F^S\}_{S \in \mathcal{N}} \right)$$

is defined as follows. The ingredients, N, T^j and $\{\pi^j(\cdot \mid t^j)\}_{t^j \in T^j}$, are already given in economy $\mathcal{E}_{pe}$. Therefore, we only need to define the choice sets C^j, the domain of strategies $T(S)$, and the feasible-strategy correspondences F^S, so that the definition of C^j enables us to use the utility functions u^j of $\mathcal{E}_{pe}$ as the utility functions of the Bayesian society. The domain $T(S)$, satisfying $\bigcup_{j \in S} T(\pi^j) \subset T(S) \subset T$, is determined by the specific context of a cooperative game played in $\mathcal{E}_{pe}$. Define

$$C^j := X^j,$$

$$F^S(\bar{x}) := \left\{ x^S : T(S) \to C^S \ \middle| \ \begin{array}{l} x^S \text{ is } \mathcal{T}^S\text{-measurable,} \\ \forall\, t : \sum_{j \in S} x^j(t) \leq \sum_{j \in S} e^j(t^j) \end{array} \right\}.$$

Thus, the utility function u^j of the Bayesian society depends only on player j's choice and a type profile. Player j's strategy here is a $\mathcal{T}^S$-measurable demand plan, $x^j : T \to \mathbf{R}_+^l$.

---

[2] We reserve the notation $X^{\{j\}}$ for the set of all logically conceivable strategies of singleton coalition $\{j\}$ in a Bayesian society; see the third paragraph following definition 2.1.2. No confusion arises, especially because we clarify the model in each of the analyses throughout this book.

Some works (e.g., Hahn and Yannelis (1997), Vohra (1999), Yazar (2001) and Ichiishi and Yamazaki (2004)) re-formulate the model so that j's strategy is a net trade plan, $z^j : t \mapsto x^j(t) - e^j(t^j)$. Demand plan x^j is $\mathcal{T}^S$-measurable iff net trade plan z^j is $\mathcal{T}^S$-measurable. Choice of demand plan versus net trade plan as a strategy affects some results (see proposition 4.1.3, theorem 10.1.1, lemma 10.1.2 and theorem 10.2.2). □

EXAMPLE 2.2.2 A *Bayesian coalition production economy*

$$\mathcal{E}_{cp} := \Big(\{X^j, T^j, u^j, e^j, \{\pi^j(\cdot \mid t^j)\}_{t^j \in T^j} \}_{j \in N}, \{Y^S\}_{S \in \mathcal{N}} \Big)$$

is an economy with l commodities, where $\{X^j, T^j, u^j, e^j, \{\pi^j(\cdot \mid t^j)\}_{t^j}\}_j$ represents the consumption sector $\mathcal{E}_{pe}$, and $\{Y^S\}_{S \in \mathcal{N}}$ represents the production sector: Correspondence $Y^S : T \to \mathbf{R}^l$ associates to each type profile t a production set $Y^S(t)$ ($\subset \mathbf{R}^l$) for coalition S. The usual sign convention is adopted for an activity $y \in Y^S(t)$, so its positive coordinates (negative coordinates, resp.) correspond to outputs (to inputs, resp.)

The associated Bayesian society

$$\Big(\{C^j, T^j, u^j, \{\pi^j(\cdot \mid t^j)\}_{t^j \in T^j}\}_{j \in N}, \{\mathbf{C}_0^S, T(S), F^S\}_{S \in \mathcal{N}} \Big)$$

is defined as follows. The ingredients, N, T^j and $\{\pi^j(\cdot \mid t^j)\}_{t^j \in T^j}$, are already given in economy $\mathcal{E}_{cp}$. Therefore, we only need to define the choice sets C^j, the domain of strategies $T(S)$, and the feasible-strategy correspondences F^S, so that the definition of C^j enables us to use the utility functions u^j of $\mathcal{E}_{cp}$ as the utility functions of the Bayesian society. The domain $T(S)$, satisfying $\bigcup_{j \in S} T(\pi^j) \subset T(S) \subset T$, is determined by the specific context of a cooperative game played in $\mathcal{E}_{cp}$. Define

$$C^j := X^j,$$

$$F^S(\bar{x}) := \left\{ x^S : T(S) \to C^S \,\middle|\, \begin{array}{l} x^S \text{ is } \mathcal{T}^S\text{-measurable,} \\ \exists\, y : T \to \mathbf{R}^l,\ \mathcal{T}^S\text{-measurable,} \\ \forall\, t \in T : y(t) \in Y^S(t), \\ \sum_{j \in S} x^j(t) \leq y(t) + \sum_{j \in S} e^j(t^j) \end{array} \right\}.$$

□

EXAMPLE 2.2.3 As a corporation grows, its internal structure evolves. One of the major characteristics of the present-day economy is emergence of *firms in multidivisional form*, (*M-form firms* in short), corporations in which several divisions (called profit centers) are operated semiautonomously; see Chandler (1962) for historical development of the M-form firms.

2: Cooperative Extension of the Bayesian Game

Each division in an M-form firm is, to a significant extent, an independent decision-maker. As an organization itself, it has its own organizational decision-making. Even if we abstract away the intraorganizational issues of a division, we still need to analyze the interorganizational issues among the divisions: As decision-units of the same corporation, these divisions talk to each other and coordinate their production activities. Total profit will then be distributed to the divisions. (The definition of profit in this example is different from the neoclassical definition of profit, in that it need not reflect the cost of resources, such as capital, that are not under the control of the divisions.)

Radner (1992) (1) formulated an interorganizational issue of the divisions as a static model of profit-center game, (2) viewed the core of the game as the equilibrium outcomes, and (3) studied its properties for several interesting cases. One of the key ingredients in Radner's formulation of an M-form firm is the distinction of *marketed commodities* and *nonmarketed commodities*; while a commodity in the former category has a price established in the market outside the firm, a commodity in the latter category has no price and is used only internally. An *intermediate nonmarketed commodity* is a commodity, not available in the market, which is supplied as an output by a division of the firm and is demanded as an input by another division.[3] A central resource allocation problem in an M-form firm then arises: A nonmarketed commodity produced or initially held by a division (say, division i) is transferred to another division (say, division j), and the two divisions have to come up with a mutually agreeable level of payment that j has to make to i in return for the use of the commodity. This problem, customarily called a *transfer payment problem*, addresses determination of prices according to a nonmarket mechanism.

We present Ichiishi and Radner's (1999) model, which introduces asymmetric information to Radner's (1992) model. It may be considered a particular instance of the Bayesian coalition production economy (example 2.2.2), but due to its focus on an information-revelation process, it has a richer structure. We will present the added structure in section 9.1 (postulates 9.1.1 and 9.1.2). For now, we only mention that the *profit center game with incomplete information* is a list of specified data of $\mathcal{D} := (\mathcal{E}_{cp}, p)$ of the coalition production economy

$$\mathcal{E}_{cp} := \left(\{\mathbf{R}^{k_m+k_n}, T^j, \text{ profit function}, r^j, \pi\}_{j \in N}, \{Y^j\}_{j \in N}\right)$$

---

[3] A division of an M-form firm, for example, produces computers and sells them in the market. The firm's product, computers, is a marketed commodity. For its production, the division needs as inputs computer chips that are produced in another division of the same firm. This chip, designed only for production of the computer, is useless for any other purpose, in particular outside the firm, so it is an intermediate nonmarketed commodity.

and a price vector of the marketed commodities $p \in \mathbf{R}_+^{k_m}$, where k_m is the number of the marketed commodities and k_n is the number of the nonmarketed commodities, and for each division j, the input-output space $\mathbf{R}^{k_m+k_n}$ is its choice set, T^j is the set of possible types, $r^j : T \to \mathbf{R}^{k_n}$ is the resource function (initial endowment vector of nonmarketed commodities), π is an *ex ante* objective probability on T, and $Y^j(\cdot)$ is the production set (so coalition S is endowed with the production set $\sum_{j \in S} Y^j(\cdot)$). Each division is risk-neutral. □

2.3 Two Approaches to Formulate Incomplete Information

In section 2.1, we followed Harsanyi's formulation of incomplete information in terms of type spaces T^j, $j \in N$, and conditional probabilities on $T^{N \setminus \{j\}}$ given type t^j, $\pi^j(\cdot \mid t^j)$, objective or subjective, for all $t^j \in T^j$ and all $j \in N$. We may consider $\pi^j(\cdot \mid t^j)$ as a conditional probability on T given t^j:

$$\pi^j(s^j, t^{N \setminus \{j\}} \mid t^j) := \begin{cases} \pi^j(t^{N \setminus \{j\}} \mid t^j) & \text{if } s^j = t^j, \\ 0 & \text{if } s^j \neq t^j. \end{cases}$$

Throughout this section, conditional probabilities in the type-space approach are understood to be defined on T.

Some authors formulate information structure in terms of a seemingly more general framework: Let $(\Omega, \mathcal{F}^j)$ be a measurable space, interpreted as follows: Set Ω is a finite state space. Family $\mathcal{F}^j$ is an algebra on Ω representing player j's *private information structure*; player j can discern states ω and ω' at the *interim* period, iff there exists $F \in \mathcal{F}^j$ such that $\omega \in F$ and $\omega' \notin F$. Notice that while $T^i \bigwedge T^j = \{\emptyset, T\}$ if $i \neq j$ in the type-space approach, the general state-space approach allows for situations in which $\mathcal{F}^i \bigwedge \mathcal{F}^j$ contains nonempty proper subsets of Ω.

Another ingredients of incomplete information are conditional probabilities, objective or subjective. Let $\mathcal{P}^j$ be the family of minimal nonempty members of $\mathcal{F}^j$; the family $\mathcal{P}^j$ constitutes a partition of Ω. Each player j holds a conditional probability on $\bigvee_{i \in N} \mathcal{F}^i$ given event F, $\pi^j(\cdot \mid F)$, for every $F \in \mathcal{P}^j$.

Just as the informational ingredients $\left\{T^j, \{\pi^j(\cdot \mid t^j)\}_{t^j \in T^j}\right\}_{j \in N}$ is common knowledge in the type-profile approach, so is the informational ingredients $\left(\Omega, \{\pi^j(\cdot \mid F)\}_{F \in \mathcal{P}^j}\right)_{j \in N}$ in the general approach.

The purpose of this section is to show that the two approaches, the type-space approach and the general state-space approach, are actually equivalent.[4] In so doing, we also show why the domain $T(S)$ of coalitional strategies, introduced in section 2.1, may be a proper subset of the type-profile space T. To the best of our knowledge, Jackson (1991) was the first to point out the equivalence.

Clearly, the type-space approach of section 2.1 is included in the general state-space approach. Indeed, given $\{T^j, \{\pi^j(\cdot \mid t^j)\}_{t^j \in T^j}\}_{j \in N}$, we simply define $\Omega := T$, $\mathcal{F}^j := \mathcal{T}^j$, and use the same conditional probabilities for Ω.

To see that the general state-space approach is included in the type-space approach, observe first that we can assume without loss of generality that $\bigvee_{i \in N} \mathcal{F}^i = 2^\Omega$. The reasoning for this observation is based on the general pattern of each player j's feasible behavior,

$$x^j(\omega) = x^j(\omega'), \text{ for all } \omega, \omega' \in F,$$
$$\text{if } F \text{ is a minimal element of } \bigvee_{i \in N} \mathcal{F}^i.$$

Indeed, given a minimal member F of $\bigvee_{i \in N} \mathcal{F}^i$, each player j finds any two states in F indiscernible at the time of taking an action, so has to schedule the same action regardless which member of F will realize (for a full discussion of the constancy of j's action on a minimal information set F, see section 3.1, in particular condition 3.1.1). Therefore, we can set $x^j(F) := x^j(\omega)$, $\omega \in F$. If $\bigvee_{i \in N} \mathcal{F}^i \neq 2^\Omega$, we can re-define Ω as the set of minimal nonempty members F of $\bigvee_{i \in N} \mathcal{F}^i$. In other words, we can re-define Ω, so that F is a "state" (an element of the newly defined Ω), and player j takes action $x^j(F)$ contingent on realization of F.

We continue our proof that the general state-space approach is included in the type-space approach. With the convention $\bigvee_{i \in N} \mathcal{F}^i = 2^\Omega$, define $T^j := \mathcal{P}^j$, and set $T := \prod_{i \in N} T^i$.

On the one hand, for each $\omega \in \Omega$, there exist uniquely $F^j \in \mathcal{P}^j$, $j \in N$, such that $\{\omega\} = \bigcap_{j \in N} F^j$ (the identity follows from $\bigvee_{i \in N} \mathcal{F}^i = 2^\Omega$), in other words, each state ω is uniquely associated with a member of T.

On the other hand, for each $t = \{F^j\}_{j \in N}$, the set $\bigcap_{j \in N} F^j$ is either empty or nonempty. In the latter case, this intersection is a singleton, due to the assumption $\bigvee_{i \in N} \mathcal{F}^i = 2^\Omega$. In other words, each type profile t is uniquely associated either with the empty set or with a member of Ω.

In the case $\bigcap_{j \in N} F^j = \emptyset$ for some $t = \{F^j\}_{j \in N}$, everybody in coalition S knows that this type profile t will never be realized, so coalition S needs not plan actions contingent on t. Therefore, $T(S) \subset T \setminus \{t\}$, in particular the domain $T(S)$ of coalitional strategies is a proper subset of the type-profile space T. The intersection $\bigcap_{h \in N} F^h$ may be empty, for example, if

---

[4]We are not asserting that the type space T and the state space Ω, both considered as measure spaces, are isomorphic.

for some $i \neq j$, $\mathcal{P}^i \cap \mathcal{P}^j$ contains a nonempty proper subset F of Ω, because

$$\bigcap_{h \in N} F^h = \emptyset, \text{ if } F^i = F \text{ and } F^j \neq F.$$

Example 2.3.1 below provides another situation in which $\bigcap_{j \in N} F^j = \emptyset$.

A probability π_Ω on Ω, *ex ante* (unconditional) or *interim* (conditional), objective or subjective, can readily be extended to a probability π_T on T. Indeed, we simply set

$$\pi_T\left((F^j)_{j \in N}\right) := \pi_\Omega \left(\bigcap_{j \in N} F^j \right), \text{ for each } (F^j)_{j \in N} \in T.$$

The members of coalition S determine the domain of their strategies, $T(S)$, subject to

$$\bigcup_{j \in S} T(\pi_T^j) \subset T(S) \subset \left\{ (F^j)_{j \in N} \in \prod_{j \in N} \mathcal{P}^j \;\middle|\; \bigcap_{j \in N} F^j \neq \emptyset \right\}.$$

Let π_Ω^i be player i's probability on Ω. We have seen that if $\bigcap_{j \in N} F^j = \emptyset$ for some $t = \{F^j\}_{j \in N}$ then $T(S)$ is a proper subset of T, and consequently the support of π_T^i is a proper subset of T. In this case, player i has a stronger information structure than T^i due to the properness of supp π_T^i (recall the discussion in the second paragraph following definition 2.1.1).

EXAMPLE 2.3.1 Suppose $N = \{1, 2\}$, $\Omega = \{a, b, c\}$, $\mathcal{P}^1 = \{\{a\}, \{b, c\}\}$, and $\mathcal{P}^2 = \{\{a, b\}, \{c\}\}$. Then, the type spaces, $T^1 := \{t_a^1, t_{bc}^1\}$ and $T^2 := \{t_c^2, t_{ab}^2\}$, can be defined as

$$t_a^1 = \{a\},\ t_{bc}^1 = \{b, c\},$$
$$t_c^2 = \{c\},\ t_{ab}^2 = \{a, b\}.$$

The type-profile space T is then identified with

	t_a^1	t_{bc}^1
t_{ab}^2	$\{a\}$	$\{b\}$
t_c^2	$\emptyset$	$\{c\}$

Let π_Ω^j be an *ex ante* probability on Ω. Then, the associated *ex ante* probability π_T^j on T satisfies

$$\text{supp } \pi_T^j \subset \left\{ (t_a^1, t_{ab}^2), (t_{bc}^1, t_c^2), (t_{bc}^1, t_{ab}^2) \right\}.$$

2: Cooperative Extension of the Bayesian Game

If $\pi_\Omega^j(\cdot \mid F^j)$, $F^j \in \mathcal{P}^j$, are *interim* probabilities on Ω, then, the associated *interim* probabilities $\pi_T^j(\cdot \mid t^j)$, $t^j \in T^j$, on T satisfies

$$\bigcup_{t^j \in T^j} \operatorname{supp} \pi_T^j(\cdot \mid t^j) \subset \{(t_a^1, t_{ab}^2), (t_{bc}^1, t_c^2), (t_{bc}^1, t_{ab}^2)\}.$$

Regardless whether a given probability on Ω is *ex ante* or *interim*, therefore, the domain $T(N)$ of strategy bundles of the grand coalition in this example satisfies

$$\bigcup_{j \in N} T(\pi_T^j) \subset T(N) \subset \{(t_a^1, t_{ab}^2), (t_{bc}^1, t_c^2), (t_{bc}^1, t_{ab}^2)\}.$$

In particular, $T(N)$ is a proper subset of T. □

EXAMPLE 2.3.2 Suppose $N = \{1, 2\}$, $\Omega = \{a, b, c\}$, $\mathcal{P}^j = \{\{a\}, \{b\}, \{c\}\}$, $j = 1, 2$ (the case in which the *interim* period and the *ex post* period are the same). Then, the type spaces $T^j := \{t_a^j, t_b^j, t_c^j\}$, $j = 1, 2$, can be defined as

$$t_a^j = \{a\}, \ t_b^j = \{b\}, \ t_c^j = \{c\}.$$

The type-profile space T is then identified with

t_c^2	∅	∅	{c}
t_b^2	∅	{b}	∅
t_a^2	{a}	∅	∅
	t_a^1	t_b^1	t_c^1

In particular, for any probability π_Ω on Ω, the support of the associated probability π_T on T is contained in the set $\{(t_a^1, t_a^2), (t_b^1, t_b^2), (t_c^1, t_c^2)\}$.

For example, if $\pi_\Omega^1(\cdot \mid F)$, $F \in \mathcal{P}^1$, are player 1's conditional probabilities on Ω given F (the only such probabilities are defined by $\pi_\Omega^1(\omega \mid F) = 1$ if $\{\omega\} = F$, $\pi_\Omega^1(\omega \mid F) = 0$ if $\{\omega\} \neq F$, for $F = \{a\}, \{b\}, \{c\}$), then the associated conditional probabilities on T given $\bar{t}_h^1$, $h = a, b, c$, are

$$\pi_T^1((t_k^1, t_l^2) \mid \bar{t}_h^1) := \begin{cases} 1 & \text{if } t_k^1 = t_l^2 = \bar{t}_h^1, \\ 0 & \text{otherwise.} \end{cases}$$

Player 1's information structure $\mathcal{T}^1(\pi^1)$ is the algebra generated by

$$\{(t_a^1, t_a^2)\}, \ \{(t_a^1, t_b^2), (t_a^1, t_c^2)\},$$
$$\{(t_b^1, t_b^2)\}, \ \{(t_b^1, t_c^2), (t_b^1, t_a^2)\},$$
$$\{(t_c^1, t_c^2)\}, \ \{(t_c^1, t_a^2), (t_c^1, t_b^2)\}.$$

The domain of strategy bundles of the grand coalition is:

$$T(N) = \left\{(t_a^1, t_a^2),\ (t_b^1, t_b^2),\ (t_c^1, t_c^2)\right\}.$$

As another example, consider the *ex ante* probabilities on Ω, given by

$$\pi_\Omega^1(\omega) = \pi_\Omega^2(\omega) = \begin{cases} 1 & \text{if } \omega = a, \\ 0 & \text{if } \omega \in \Omega \setminus \{a\}, \end{cases}$$

in other words, both players believe *ex ante* that state a will occur with probability 1. Then, the associated *ex ante* probabilities π_T^j, $j = 1, 2$, assign probability 1 to the type profile (t_a^1, t_a^2), so the grand coalition chooses the domain $T(N)$ of their *ex ante* strategies, subject to

$$\left\{(t_a^1, t_a^2)\right\} \subset T(N) \subset \left\{(t_a^1, t_a^2),\ (t_b^1, t_b^2),\ (t_c^1, t_c^2)\right\}.$$

□

REMARK 2.3.3 One might be tempted to define the domain of coalitional strategies only by looking at the logical aspect of the definition of type profiles:

$$T(S) := \left\{ (F^j)_{j \in N} \in \prod_{j \in N} \mathcal{P}^j \ \middle|\ \bigcap_{j \in N} F^j \neq \emptyset \right\}.$$

This definition, while being appropriate in many cases, overlooks the probabilistic aspect. Our more general treatment of $T(S)$, on the other hand, allows for the scenario that the members of coalition S plan their actions only for those type profiles with a positive probability, even if they hold merely subjective probabilities. □

Chapter 3

Measurability as Feasibility of Individual Actions

This chapter and next chapter present two economically meaningful conditions that the strategies of a Bayesian society (definition 2.1.3) have to satisfy in the presence of differential information: measurability with respect to the available information structure, and Bayesian incentive compatibility. We first discuss the issue of measurability.

3.1 General Case

Suppose that the grand coalition in a Bayesian society

$$\mathcal{S} := \left(\{C^j, T^j, u^j, \{\pi^j(\cdot \mid t^j)\}_{t^j \in T^j}\}_{j \in N}, \{\mathbf{C}_0^S, T(S), F^S\}_{S \in \mathcal{N}} \right)$$

is entertaining a strategy bundle $\bar{x}$, but that the members of coalition S are contemplating to defect and to take their own strategy bundle $x^S : T(S) \to C^S$, *ex ante* or *interim*.

Suppose the members of S know that a communication system $\{\mathcal{A}^j\}_{j \in S}$ will be available to them at the time of strategy execution, that is, when each member j will make choice according to his strategy x^j. If t is the true type profile, denoting by $\mathcal{A}^j(t)$ the minimal element of $\mathcal{A}^j$ that contains t, player j will know at the time of action that the event $\mathcal{A}^j(t)$ has occurred, but he will not know which specific state in the event $\mathcal{A}^j(t)$ has actually realized. Since he cannot distinguish the states in $\mathcal{A}^j(t)$, he cannot take different actions for any two states in $A^j(t)$. This means that his strategy x^j has to be constant on $A^j(t)$.[1] In other words, his strategy has to be $\mathcal{A}^j$-measurable.

Within the general equilibrium framework, Radner (1968) proposed the measurability condition with respect to an available information structure as a feasibility requirement on individual actions. By adopting this feasibility condition for a coalitional framework, we obtain the following condition.

CONDITION 3.1.1 (Radner, 1968) Suppose that the grand coalition is entertaining a strategy bundle $\bar{x}$, but that the members of coalition S are contemplating to defect and to take their own strategy bundle. Suppose the members of S know that a communication system $\{\mathcal{A}^j\}_{j \in S}$ will be available to them at the time of strategy execution. They can take only those strategies $x^S \in F^S(\bar{x})$ such that x^j is $\mathcal{A}^j$-measurable for every $j \in S$.

3.2 Private Information Case, Fully Pooled Information Case

The *private information case* of a Bayesian society $\mathcal{S}$ is defined as the situation in which when the members execute a strategy bundle, member j has only his private information structure $\mathcal{T}^j$, so knows only his true type t^j and the *interim* probability $\pi^j(\cdot \mid t^j)$ on the others' types. In this case the measurability condition 3.1.1 becomes the *private measurability condition* in that each player j's strategy be $\mathcal{T}^j$-measurable. Notice that a strategy $x^j : T(S) \to C^j$ is $\mathcal{T}^j$-measurable, iff it is a function only of t^j. We may, therefore, safely write $x^j(t^j)$ (instead of $x^j(t)$) in the private information case.

Yannelis (1991) re-emphasized the significance of the measurability condition by introducing the private measurability to the core analysis of the

---

[1] A conference on economic theory is held at a respectable university on the outskirts of Istanbul, Turkey, and many theorists from the world over participate in it. One day during the conference, all the American participants have disappeared from the conference site. The rest of the conference participants are told that the Americans are in the old town of Istanbul, enjoying sightseeing. The precise whereabouts of the American group is the Americans' private information and this information has not been made public; they may be at Topkapi or at the Blue Mosque. A Japanese participant, not knowing how to enjoy life, stays at the conference; he wants to have coffee or tea during the session break. He cannot make his choice of coffee versus tea contingent upon the American group's location as he does not know it; it is impossible for him, for example, to choose to drink coffee based on the fact that the Americans are visiting Topkapi and to choose tea based on the fact that they are visiting the Blue Mosque. He can plan to drink coffee regardless whether the Americans are at Topkapi or at the Blue Mosque, or can plan to drink tea regardless whether the Americans are at Topkapi or at the Blue Mosque.

Bayesian pure exchange economy. For the private information case, define the correspondences $F'^S : X \to X^S$ by

$$F'^S(\bar{x}) := \{x^S \in F^S(\bar{x}) \mid \forall\, j \in S : x^j \text{ is } \mathcal{T}^j\text{-measurable.}\}.$$

The *fully pooled information case* is defined as the situation in which when the members execute the strategy bundle x^S, every member j has the pooled information structure $\mathcal{T}^S$. As soon as we go beyond the private information case, it is highly desirable to explain how agents come to pool or share their private information. This issue will be picked up in chapters 4 and 9.

Some authors (see, e.g., Vohra (1999) or Forges, Minelli, and Vohra (2002)) emphasize that the private measurability requirement is too stringent as a feasibility condition. However, we would like to note, as was done by Radner (1968), that the measurability with respect to the information structure of an individual at the time of his action is a fundamental feasibility requirement. Of course, this measurability requirement need not be the private measurability. But as we just remarked, if the measurability requirement is to take account of information revealed by other players, it is highly desirable to model information revelation explicitly as in the rational expectations model (Radner, 1979). Without such an explicit modeling of information revelation, the private measurability requirement seems highly plausible.

Chapter 4

Bayesian Incentive Compatibility as Feasibility of Execution of Contracts

The next feasibility condition pertains to the feasibility of execution of strategy bundles viewed as "contracts" made within a coalition. In order for a strategy bundle to be agreed upon by the players, each player must feel certain that the strategy bundle will be executed in exact accordance with its terms. This feasibility condition is formulated as the Bayesian incentive compatibility. We start discussions of this condition for the private information case in section 4.1. We will discuss the same condition, modified for the other cases, in sections 4.2 and 4.3.

4.1 Private Information Case

Consider a Bayesian society,

$$\mathcal{S} := \left(\{C^j, T^j, u^j, \{\pi^j(\cdot \mid t^j)\}_{t^j \in T^j}\}_{j \in N}, \{\mathbf{C}_0^S, T(S), F^S\}_{S \in \mathcal{N}} \right),$$

in the private information case. Recall that a strategy $x^j : T(S) \to C^j$ is $\mathcal{T}^j$-measurable, iff it is a function only of t^j. To execute a $\mathcal{T}^j$-measurable strategy $x^j : T(S) \to C^j$ as a member of coalition S, player j of type $\bar{t}^j$ needs to take action $x^j(\bar{t}^j)$. We sometimes say throughout the present book,

"Player j represents or poses as type $\bar{t}^j$." In the traditional terminology of the mechanism theory, "Player j reports his type $\bar{t}^j$." Unlike most of the noncooperative theory, however, *the present cooperative theory does not suppose the existence of a mediator who would receive reports from the players*; in fact no player needs to report his type to any other players or third person who is outside the game. What we mean by this statement is that he simply makes the choice $c^j := x^j(\bar{t}^j) \in C^j$. The other players may observe the action c^j, but does not have to be told about the type $\bar{t}^j$. This interpretation is particularly important when the function x^j is not 1-1 (i.e., when x^j is not fully information-revealing), since j's action c^j does not provide the information to his colleagues as to which of the types $\left(x^j\right)^{-1}(c^j)$ is his true type. In the private information case of the one-shot game, player j can make any choice from the range of function x^j and can avoid being caught.

The members of a coalition agree on a strategy bundle, in order to plan a best choice bundle preparing for every contingency. It is essential, therefore, that choices are later made as scheduled. But the private information case in particular creates the incentive for players to misrepresent their true types after a strategy is chosen, which would prevent realization of the planned result, thereby failing to fulfill the purpose of coalition formation. If members of coalition S foresee at the outset that a particular strategy bundle may later induce such misrepresentation, they will not agree to such a bundle. Therefore, the feasible-strategy set is further restricted to those strategies that are Bayesian incentive-compatible in the sense of d'Aspremont and Gérard-Varet (1979). We elaborate on this idea in this section.

Suppose that the grand coalition is entertaining a strategy bundle $\bar{x}$, but that the members of coalition S are contemplating to defect and to take their own private measurable strategy bundle $x^S : T(S) \to C^S$ after defection. Let j be any member of S, whose true type is $\bar{t}^j$. If he makes a choice according to the agreement, his conditional expected utility given his true type is

$$Eu^j(x^S, \bar{x}^{N\setminus S} \mid \bar{t}^j)$$
$$:= \sum_{t^{N\setminus\{j\}} \in T^{N\setminus\{j\}}} u^j\left(x^j(\bar{t}^j), x^{S\setminus\{j\}}(t^{S\setminus\{j\}}), \bar{x}^{N\setminus S}(t^{N\setminus S}), (\bar{t}^j, t^{N\setminus\{j\}})\right)$$
$$\times \pi^j(t^{N\setminus\{j\}} \mid \bar{t}^j).$$

If on the other hand he makes choice $c^j \in x^j(T(S)) \setminus \{x^j(\bar{t}^j)\}$ contrary to

4: Bayesian Incentive Compatibility Requirement

the agreement,[1] his conditional expected utility given his true type is

$$Eu^j(c^j, x^{S\setminus\{j\}}, \bar{x}^{N\setminus S} \mid \bar{t}^j)$$
$$:= \sum_{t^{N\setminus\{j\}} \in T^{N\setminus\{j\}}} u^j\left(c^j, x^{S\setminus\{j\}}(t^{S\setminus\{j\}}), \bar{x}^{N\setminus S}(t^{N\setminus S}), (\bar{t}^j, t^{N\setminus\{j\}})\right)$$
$$\times \pi^j(t^{N\setminus\{j\}} \mid \bar{t}^j).$$

Roughly stated, his colleagues $S \setminus \{j\}$ cannot catch this betraying act in the private information case, being led to believe that j's true type were in the event $(x^j)^{-1}(c^j)$. In the present context, the *Bayesian incentive compatibility* says that strategy bundle x^S is designed so that nobody has the incentive to act contrary to the promised strategy bundle x^S, in other words, $Eu^j(x^S, \bar{x}^{N\setminus S} \mid \bar{t}^j) \geq Eu^j(c^j, x^{S\setminus\{j\}}, \bar{x}^{N\setminus S} \mid \bar{t}^j)$.

One subtle point is that player j may not be able to choose an arbitrary action $c^j \in x^j(T(S))$ even in the private information case, because some colleague $i \in S \setminus \{j\}$, having his own *interim* probability, may detect j's betraying act and accuse j of breach of contract. In the rest of this section we adopt the convention that *interim* probabilities $\pi^i(\cdot \mid \bar{t}^i)$ are defined on $\{\bar{t}^i\} \times T^{N\setminus\{i\}}$ rather than on $T^{N\setminus\{i\}}$; see the first paragraph of section 2.3. Let $\bar{t}$ be the true type profile. Player i knows for sure that the true type profile cannot be t if $\pi^i(t \mid \bar{t}^i) = 0$, so player j cannot take an action c^j if $\pi^i\left((x^j)^{-1}(c^j) \mid \bar{t}^i\right) = 0$. In fact, all of j's colleagues $S \setminus \{j\}$ may pool their information and deduce that the true type profile cannot be t if $t \notin \bigcap_{i \in S\setminus\{j\}} \text{supp } \pi^i(\cdot \mid \bar{t}^i)$. Thus, when the true type profile is $\bar{t}$, player j can take an action c^j without being caught, iff

$$c^j \in x^j\left(\bigcap_{i \in S\setminus\{j\}} \text{supp } \pi^i(\cdot \mid \bar{t}^i)\right).$$

CONDITION 4.1.1 (d'Aspremont and Gérard-Varet, 1979) Suppose that the grand coalition is entertaining a strategy bundle $\bar{x}$, but that the members of coalition S are contemplating to defect and to take their own strategy bundle after defection. In the private information case, members of S agree only on those strategies $x^S \in F'^S(\bar{x})$ that are *Bayesian incentive-compatible*, that is,

$$\forall j \in S : \forall \bar{t} \in T(S) : \forall c^j \in x^j\left(\bigcap_{i \in S\setminus\{j\}} \text{supp } \pi^i(\cdot \mid \bar{t}^i)\right) :$$
$$Eu^j(x^S, \bar{x}^{N\setminus S} \mid \bar{t}^j) \geq Eu^j(c^j, x^{S\setminus\{j\}}, \bar{x}^{N\setminus S} \mid \bar{t}^j).$$

---
[1] Recall the present emphasis that players take actions (make choices) rather than make reports; in fact they do not have to report their types to anybody.

We make two remarks on condition 4.1.1. The first remark is that, in the private information case, player j does not precisely know the true type profile $\bar{t}$, let alone the set supp $\pi^i(\cdot \mid \bar{t}^i)$, at the time of his action, but as long as the strategy bundle x^S satisfies condition 4.1.1, he knows that dishonest action will either be detected (hence his receiving severe penalty from his organization) or else will give rise to an *interim* utility level no greater than the *interim* utility level of honest action.

The second remark is that, in order to be consistent with our interpretation of the model that a member does not have to report his type but merely has to make a choice, one might argue that condition 4.1.1 should be modified to:

$$\forall\, j \in S : \forall\, \bar{t} \in T(S) : \forall\, c^j \in \bigcap_{i \in S\setminus\{j\}} x^j\left(\text{supp } \pi^i(\cdot \mid \bar{t}^i)\right) :$$
$$Eu^j(x^S, \bar{x}^{N\setminus S} \mid \bar{t}^j) \geq Eu^j(c^j, x^{S\setminus\{j\}}, \bar{x}^{N\setminus S} \mid \bar{t}^j).$$

This condition is stronger than condition 4.1.1, due to the fact

$$\bigcap_{i \in S\setminus\{j\}} x^j\left(\text{supp } \pi^i(\cdot \mid \bar{t}^i)\right) \supset x^j\left(\bigcap_{i \in S\setminus\{j\}} \text{supp } \pi^i(\cdot \mid \bar{t}^i)\right).$$

Condition 4.1.1 on the other hand is based on player j's very conservative attitude that he never risks being caught, preparing even for the situations in which his colleagues explicitely pool their information and see that the event $\bigcap_{i \in S\setminus\{j\}} \text{supp } \pi^i(\cdot \mid \bar{t}^i)$ has been realized.

To see the effect of the *interim* probabilities on the Bayesian incentive compatibility condition, we will see two extreme cases. Consider first the case of extreme asymmetry in information: $T(\pi^j) = T$ for all $j \in N$. Then, $T(S) = T$ for all $S \in \mathcal{N}$. Condition 4.1.1 in this case is simplified as

$$\forall\, j \in S : \forall\, \bar{t}^j \in T^j : \forall\, c^j \in x^j(T) :$$
$$Eu^j(x^S, \bar{x}^{N\setminus S} \mid \bar{t}^j) \geq Eu^j(c^j, x^{S\setminus\{j\}}, \bar{x}^{N\setminus S} \mid \bar{t}^j).$$

Next, consider the case in example 2.3.2 in which player i is endowed with the *interim* probabilities $\pi^i(\cdot \mid t^i_h)$, $h = a, b, c$. Then,

$$T(N) = \{(t^1_a, t^2_a),\ (t^1_b, t^2_b),\ (t^1_c, t^2_c)\},$$

and each set supp $\pi^i(\cdot \mid t^i_h)$ is singleton $\{(t^1_h, t^2_h)\}$, $h = a, b, c$, so any private measurable, feasible strategy automatically satisfies condition 4.1.1.

In accordance with d'Aspremont and Gérard-Varet's original formulation, the Bayesian incentive compatibility can also be stated in the following

4: Bayesian Incentive Compatibility Requirement

way. For simplicity of the argument, we assume the extreme asymmetry of information, $T(\pi^j) = T$ for all $j \in N$. Recall the definition of a Bayesian game, $\{C^j, T^j, u^j, \{\pi^j(\cdot \mid t^j)\}_{t^j \in T^j}\}_{j \in N}$ (definition 2.1.1). After the members of coalition S have agreed on a strategy bundle $x^S \in F'^S(\bar{x})$, they play the Bayesian game with the player set S,

$$\mathcal{BG}(x^S, \bar{x}^{N\setminus S}) := \left\{T^j, T^j, U^j, \{\pi^j_{T^S}(\cdot \mid t^j)\}_{t^j \in T^j}\right\}_{j \in S},$$

where
$$U^j(t^S, \bar{t}^S) := Eu^j(x^S(t^S), \bar{x}^{N\setminus S} \mid \bar{t}^S)$$

in which the expectation is taken with respect to the outsiders' type profiles $t^{N\setminus S}$. The probability $\pi^j_{T^S}(\cdot \mid t^j)$ is the marginal probability of $\pi^j(\cdot \mid t^j)$ on T^S. Player j's choice space is his type space T^j, so that his strategy is a *pretension function* $\sigma^j : T^j \to T^j$, specifying for each possible true type $\bar{t}^j$ his reported type $\sigma^j(\bar{t}^j)$. The value $U^j(t^S, \bar{t}^S)$ is player j's "*ex post*" utility level in Bayesian game $\mathcal{BG}(x^S, \bar{x}^{N\setminus S})$, when the players S's true type profile is $\bar{t}^S$, but these players pretend that their types are t^S. The members of S are assuming that the outsiders $N\setminus S$ do not pretend but truthfully execute the strategy bundle $\bar{x}^{N\setminus S}$. When the members choose a strategy bundle $\{\sigma^j\}_{j \in S}$ and their true type profile is $\bar{t}^S$, each member j's utility level is $U^j(\sigma^S(\bar{t}^S), \bar{t}^S) = Eu^j((x^i(\sigma^i(\bar{t}^i)))_{i \in S}, \bar{x}^{N\setminus S} \mid \bar{t}^S))$. Of course, at the time of playing this Bayesian game, each player $j \in S$ has only his private information, $\bar{t}^j$, so his *interim* utility is the conditional expected utility given $\bar{t}^j$,

$$Eu^j(x^S \circ \sigma^S, \bar{x}^{N\setminus S} \mid \bar{t}^j) := \sum_{t^S \in T^S} U^j(\sigma^S(t^S), t^S) \pi^j_{T^S}(t^S \mid \bar{t}^j).$$

A *Bayesian equilibrium* is a strategy bundle σ^{*S} such that for any possible true type profile $\bar{t}^S$, the choice bundle $\sigma^{*S}(\bar{t}^S)$ is a Nash equilibrium:

$$\forall \bar{t}^S \in T^S : \forall j \in S : \forall t^j \in T^j :$$
$$Eu^j(x^S \circ \sigma^{*S}, \bar{x}^{N\setminus S} \mid \bar{t}^j) \geq Eu^j(x^j(t^j), x^{S\setminus\{j\}} \circ \sigma^{*S\setminus\{j\}}, \bar{x}^{N\setminus S} \mid \bar{t}^j),$$

where $x^j(t^j)$ is the constant function, $t'^j \mapsto x^j(t^j)$. *Bayesian incentive compatibility* says that the identity function from T^S to T^S is a Bayesian equilibrium. Thus, player j finds it to his advantage to make an honest report assuming that the others $N \setminus \{j\}$ are also making honest reports.

Bayesian incentive compatibility had been used in Myerson's (1984) study of the λ-transfer value in the context of incomplete information. Ichiishi and Idzik (1996) introduced Bayesian incentive compatibility to the Bayesian core analysis (or more generally, to the Bayesian strong equilibrium analysis) of the extreme case, $T(\pi^j) = T$ for all $j \in N$.[2]

For the private information case, define the correspondences $\hat{F}^S : X \to X^S$ by

$$\hat{F}^S(\bar{x}) := \{x^S \in F'^S(\bar{x}) \mid x^S \text{ is Bayesian incentive-compatible.}\}.$$

In many interesting private information cases, members of the set $\hat{F}^S(\bar{x})$ are abundant; any constant function, for example, is a member of $\hat{F}^S(\bar{x})$ provided it is a member of $F^S(\bar{x})$.

A strategy bundle x^S is called *strictly Bayesian incentive-compatible* relative to the grand coalition's strategy bundle $\bar{x}$, if the Bayesian incentive compatibility condition, expressed by the weak inequalities, is satisfied with strict inequalities,

$$\forall j \in S : \forall \bar{t} \in T(S) : \left(\forall \tilde{t} \in \bigcap_{i \in S \setminus \{j\}} \operatorname{supp} \pi^i(\cdot \mid \bar{t}^i) : \bar{t}^j \neq \tilde{t}^j \right) :$$
$$Eu^j(x^S, \bar{x}^{N \setminus S} \mid \bar{t}^j) > Eu^j(x^j(\tilde{t}^j), x^{S \setminus \{j\}}, \bar{x}^{N \setminus S} \mid \bar{t}^j).$$

A sufficient condition for the existence of a strictly Bayesian incentive-compatible strategy bundle can be found in social choice theory:

LEMMA 4.1.2 (Abreu and Matsushima, 1992) *Let* $(\{C^j, T^j, u^j\}_{j \in N},$ $\{\mathbf{C}_0^S, T(S), F^S\}_{S \in \mathcal{N}}, \pi)$ *be a Bayesian society such that* $T(S) = T$. *Assume for each player* $j \in N$ *that his choice set* C^j *is a convex subset of a vector space, his utility function* u^j *is defined on* $C^j \times T^j$, *and the function* $u^j(\cdot, t^j)$ *is affinely linear on* C^j *for every* $t^j \in T^j$. *Assume also that there exists a finite subset* C_f^j *of* C^j *such that*

$$(\forall t^j, t'^j \in T^j : t^j \neq t'^j) : \exists c^j, c'^j \in C_f^j :$$
$$u^j(c^j, t^j) > u^j(c'^j, t^j) \text{ and } u^j(c'^j, t'^j) > u^j(c^j, t'^j).$$

Then, there exists a 1-1 function $x^j : T^j \to C^j$ *such that*

$$(\forall \bar{t}^j, \tilde{t}^j \in T^j : \bar{t}^j \neq \tilde{t}^j) : u^j(x^j(\bar{t}^j), \bar{t}^j) > u^j(x^j(\tilde{t}^j), \bar{t}^j).$$

Proof We will construct the required strategy x^j whose image is in the convex hull of C_f^j. For each $t^j \in T^j$, fix a linear order on C_f^j,

$$c(t^j, 1), \ c(t^j, 2), \ \ldots \ c(t^j, K),$$

---
[2] The first draft of Ichiishi and Idzik (1996) had been circulated since the summer of 1991.

4: Bayesian Incentive Compatibility Requirement

so that

$$u^j\left(c(t^j,1),t^j\right) \geq u^j\left(c(t^j,2),t^j\right) \geq \ldots \geq u^j\left(c(t^j,K),t^j\right),$$

where $K := \#C_f^j$. Different orders are associated with different types. Choose $\alpha \in \mathbf{R}_+^K$ such that

$$\alpha_1 > \alpha_2 > \ldots > \alpha_K,$$
$$\alpha_1 + \alpha_2 + \ldots + \alpha_K = 1.$$

Define for each t^j,

$$x^j(t^j) := \sum_{k=1}^K \alpha_k c^j(t^j, k).$$

In defining $x^j(t^j)$ as the weighted sum of the choices $c(t^j, k)$, $k = 1, 2, \ldots K$, the choice $c(t^j, 1)$ that yields the highest utility in C_f^j is given the highest weight α_1, the choice $c(t^j, 2)$ that yields the highest utility in $C_f^j \setminus \{c(t^j, 1)\}$ is given the second highest weight α_2, etc. Therefore, $u^j\left(x^j(t^j), t^j\right) \geq u^j\left(x^j(t'^j), t^j\right)$. This inequality is strict if $t^j \neq t'^j$, since there are at least two choices whose rankings are opposite according to the orders associated with t^j and t'^j.

The plan x^j is 1-1 on T^j. Indeed, if $t^j \neq t'^j$, then $u^j(x^j(t^j), t^j) > u^j(x^j(t'^j), t^j)$, so $x^j(t^j) \neq x^j(t'^j)$. □

The dependence of j's utility function u^j only on j's own choice and type in lemma 4.1.2 may be called the *no-externality case*, but the other players still influence j through the feasible-strategy correspondences, F^S, $S \ni j$. For application of the above lemma to strict Bayesian incentive compatibility, it suffices to notice that $Eu^j(x \mid t^j) = u^j(x^j(t^j), t^j)$ in the present no-externality case.

Extension of lemma 4.1.2 to situations in which $T(S)$ is a proper subset of T is straightforward, and is left to the reader.

Hahn and Yannelis (1997) noted that for the Bayesian pure exchange economy in the private information case, the private measurability condition (condition 3.1.1 for the null communication system) implies the Bayesian incentive compatibility condition (condition 4.1.1), provided that the consumers' strategies are net trade plans:

PROPOSITION 4.1.3 (Hahn and Yannelis, 1997) *Let $\mathcal{E}_{pe}$ be the Bayesian pure exchange economy in the private information case, in which each player j's strategy is his net trade plan and the coalitional feasibility is defined by the equality of supply and demand within the each coalition. Then the private measurability condition implies the Bayesian incentive compatibility condition.*

Proof Let $z^S : T(S) \to \mathbf{R}^{l \cdot \#S}$ be a feasible strategy bundle of coalition S which satisfies the private measurability condition. Then, each z^j is a function of t^j, and

$$\forall \, j \in S : \forall \, t \in T(S) : z^j(t^j) = - \sum_{i \in S \setminus \{j\}} z^i(t^i).$$

Let $\bar{t} \in T(S)$ be the true type profile, and choose any consumer $j \in S$ and any $t \in \bigcap_{i \in S \setminus \{j\}} \operatorname{supp} \pi^i(\cdot \mid \bar{t}^i) \subset \bigcap_{i \in S \setminus \{j\}} \{\bar{t}^i\} \times T^{N \setminus \{i\}}$. Then $t^i = \bar{t}^i$ for all $i \in S \setminus \{j\}$. The attainability has to be satisfied for the two type profiles $t = (t^j, \bar{t}^{S \setminus \{j\}}, t^{N \setminus S})$ and $\bar{t}$, so

$$z^j(t^j) = - \sum_{i \in S \setminus \{j\}} z^i(\bar{t}^i) = z^j(\bar{t}^j).$$

Then,

$$\begin{aligned}
Eu^j(z^j(t^j) + e^j \mid \bar{t}^j) &= Eu^j(z^j(t^j) + e^j(\bar{t}^j) \mid \bar{t}^j) \\
&= Eu^j(z^j(\bar{t}^j) + e^j(\bar{t}^j) \mid \bar{t}^j) \\
&= Eu^j(z^j + e^j \mid \bar{t}^j),
\end{aligned}$$

so strategy z^j is Bayesian incentive-compatible. □

Notice that the feasibility of strategies in this proposition is given by the equality (of the total demand and the total supply), but the feasibility in example 2.2.1 is given by the weak inequality. It turns out that under the weak monotonicity assumption on the preference relations, if some measurable strategy bundle satisfies the market clearance condition with weak inequality, then there is a larger measurable strategy bundle which satisfies the market clearance condition with equality (see lemma 9.2.10).

This proposition is no longer valid if a demand plan is used as a strategy, as the following example 4.1.4 shows. The proposition is not valid either in the general model of Bayesian society $\mathcal{S}$.

EXAMPLE 4.1.4 A Bayesian pure exchange economy with one commodity ($l = 1$), in which consumer j's type space consists of two elements, $T^j = \{a^j, b^j\}$, his utility function is the identity function, $u^j(c^j, t) = c^j$, and his initial endowment is given by

$$e^j(t^j) = \begin{cases} 1, & \text{if } t^j = a^j, \\ 2, & \text{if } t^j = b^j. \end{cases}$$

Probabilities, *ex ante* or *interim*, are given in such a way that $T(\pi^j) = T$ for all $j \in N$, in particular, $T(S) = T$ for all $S \in \mathcal{N}$. Then, the initial endowment bundle e is attainable in N with equality and satisfies measurability. But

$$Eu^j(e^j(b^j) \mid a^j) = 2 > 1 = Eu^j(e^j \mid a^j),$$

so strategy e^j is not Bayesian incentive-compatible. □

One might consider players who are bold enough to believe that they will not be caught for a dishonest action as long as the action is planned for a type profile which at least one colleague, without pooling information, thinks will be realized with a positive probability. The members of a coalition with such bold colleagues have to design a strategy so that nobody will benefit from dishonest action, bold or conservative. The Bayesian incentive compatibility condition (condition 4.1.1) is strengthened in this case to:

$$\forall\, j \in S : \forall\, \bar{t} \in T(S) : \forall\, c^j \in x^j \left(\bigcup_{i \in S \setminus \{j\}} \operatorname{supp} \pi^i(\cdot \mid \bar{t}^i) \right) :$$
$$Eu^j(x^S, \bar{x}^{N \setminus S} \mid \bar{t}^j) \geq Eu^j(c^j, x^{S \setminus \{j\}}, \bar{x}^{N \setminus S} \mid \bar{t}^j).$$

Analysis of implications of this strengthened Bayesian incentive compatibility condition is outside the scope of this book. We only note in the following example 4.1.5 that proposition 4.1.3 is false for this strengthened Bayesian incentive compatibility condition.

EXAMPLE 4.1.5 Let

$$\mathcal{E}_{pe} := \left\{ \mathbf{R}_+, T^j, u^j, e^j, \{\pi^j(\cdot \mid t^j)\}_{t^j \in T^j} \right\}_{j=1,2,3}$$

be the Bayesian pure exchange economy with one commodity ($l = 1$) and three consumers ($N = \{1, 2, 3\}$) in the private information case, in which each player j's strategy is his net trade plan and the coalitional feasibility is defined by the equality of supply and demand within the each coalition. We will construct a specific example of $\mathcal{E}_{pe}$ for which the private measurability condition does not imply the Bayesian incentive compatibility condition strengthened as in the preceding paragraph.

The type spaces are given as

$$T^1 = \{t_a^1, t_{bc}^1\}, \ T^2 = \{t_{ab}^2, t_c^2\}, \ T^3 = \{t_a^3, t_b^3, t_c^3\}.$$

Each consumer j holds his *interim* probability given t_h^j, $\pi^j(\cdot \mid t_h^j)$, such that

$$\operatorname{supp} \pi^1(\cdot \mid t_h^1) = \begin{cases} \{(t_a^1, t_{ab}^2, t_a^3)\}, & \text{if } h = a, \\ \{(t_{bc}^1, t_{ab}^2, t_b^3), (t_{bc}^1, t_c^2, t_c^3)\}, & \text{if } h = bc, \end{cases}$$

$$\text{supp } \pi^2(\cdot \mid t_h^2) = \begin{cases} \{(t_a^1, t_{ab}^2, t_a^3), (t_{bc}^1, t_{ab}^2, t_b^3)\}, & \text{if } h = ab, \\ \{(t_{bc}^1, t_c^2, t_c^3)\}, & \text{if } h = c, \end{cases}$$

$$\text{supp } \pi^3(\cdot \mid t_h^3) = \begin{cases} \{(t_a^1, t_{ab}^2, t_a^3)\}, & \text{if } h = a, \\ \{(t_{bc}^1, t_{ab}^2, t_b^3)\}, & \text{if } h = b, \\ \{(t_{bc}^1, t_c^2, t_c^3)\}, & \text{if } h = c, \end{cases}$$

$$T(S) = \{(t_a^1, t_{ab}^2, t_a^3), (t_{bc}^1, t_{ab}^2, t_b^3), (t_{bc}^1, t_c^2, t_c^3)\}.$$

[This is derived as in section 2.3 from the following example of the general approach to incomplete information,

$$\begin{aligned}
\Omega &= \{a, b, c\}, \\
\mathcal{P}^1 &= \{\{a\}, \{b, c\}\} =: \{t_a^1, t_{bc}^1\}, \\
\mathcal{P}^2 &= \{\{a, b\}, \{c\}\} =: \{t_{ab}^2, t_c^2\}, \\
\mathcal{P}^3 &= \{\{a\}, \{b\}, \{c\}\} =: \{t_a^3, t_b^3, t_c^3\},
\end{aligned}$$

in which each *interim* probability $\pi_\Omega^j(\cdot \mid P)$ given $P \in \mathcal{P}^j$ is strictly positive on P.] Each consumer j's utility function is given as the identity function,

$$u^j(c^j, t) = c^j,$$

and his initial endowment function is a constant function,

$$e^j(t^j) = 2, \text{ for all } t^j \in T^j.$$

Now, consider the following strategy bundle (excess demand plans):

$$z^1(t_h^1) = \begin{cases} 2 & \text{if } h = a, \\ 1 & \text{if } h = bc, \end{cases}$$

$$z^2(t_h^2) = \begin{cases} -1 & \text{if } h = ab, \\ -2 & \text{if } h = c, \end{cases}$$

$$z^3(t_h^3) = \begin{cases} -1 & \text{if } h = a, \\ 0 & \text{if } h = b, \\ 1 & \text{if } h = c, \end{cases}$$

It satisfies the private measurability condition by definition, and it is easy to verify $\sum_{j \in N} z^j(t) = 0$ for all $t \in T(N)$.

However, this strategy bundle does not satisfy the strengthened Bayesian incentive compatibility condition. Indeed, suppose $\bar{t} := (t_{bc}^1, t_{ab}^2, t_b^3)$ is the true type profile. Then,

$$\bigcup_{i \in N \setminus \{3\}} \text{supp } \pi^i(\cdot \mid \bar{t}) = \{(t_{bc}^1, t_{ab}^2, t_b^3), (t_{bc}^1, t_c^2, t_c^3), (t_a^1, t_{ab}^2, t_a^3)\},$$

so consumer 3 has the incentive to choose $z^3(t_c^3) = 1$, although he is supposed to choose $z^3(t_b^3) = 0$.

Notice that although this strategy bundle is not constant functions, it satisfies condition 4.1.1 (Bayesian incentive compatibility), as asserted in proposition 4.1.3. For a direct proof of this fact, we simply verify that

$$\forall j \in N : \forall \bar{t} \in T(N) : \bigcap_{i \in N \setminus \{j\}} \text{supp } \pi^i(\cdot \mid \bar{t}^i) = \{\bar{t}\}.$$

$\square$

So far, we have seen formulations of Bayesian incentive compatibility in the private information case within the framework of one-shot model. More generally, Bayesian incentive compatibility reflects the information system available at the time of action (strategy execution). Section 9.1 and the appendix to chapter 9 present how the definition is modified for a particular two-*interim*-period model. We will also present another formulation of Bayesian incentive compatibility in chapter 11, which is useful for analysis of a large economy.

We conclude this section by presenting warning against mechanical application of the revelation principle; the principle was established for the principal-agent theory. [The readers who are not interested in the revelation principle in our context can skip the rest of this section.]

First, review of the revelation principle in the principal-agent theory is in order. There is one uninformed principal (the first mover – the Stackelberg leader), and a finite set N of agents who have private information (the second movers – the Stackelberg followers). Let M^j be agent j's message space (in a special case, the space M^j may be the type space, T^j), and set $M := \prod_{j \in N} M^j$. Let Z be an outcome space. As the first mover, the principal designs a mechanism $g : M \to Z$ which specifies an outcome to each message profile, and offers it to the agents. Each agent then decides to accept or reject it. If all agents accept it, each agent j sends a message to the principal; he can condition his message on his private information, so his strategy is a function, $\sigma^j : T^j \to M^j$, which means that agent j sends message $\sigma^j(t^j)$ if his true type is t^j. Denoting by $u^j : Z \times T^j \to \mathbf{R}$ agent j's type-dependent von Neumann-Morgenstern utility function defined on the outcome space, the agents play the Bayesian game $\{M^j, T^j, u^j \circ g, \{\pi^j(\cdot \mid t^j)\}_{t^j \in T^j}\}_{j \in N}$. A strategy bundle σ gives rise to agent j's *interim* conditional expected utility $Eu^j(g(\sigma) \mid \bar{t}^j)$ given his type $\bar{t}^j$:

$$Eu^j(g(\sigma) \mid \bar{t}^j) := \sum_{t \in T} \pi^j(\{t\} \mid \bar{t}^j) u^j(g(\sigma(t)), \bar{t}^j).$$

A *Bayesian equilibrium* of this game is a strategy bundle σ^* such that for all player $j \in N$ and all his possible type $t^j \in T^j$, his strategy σ^{*j} gives the best response (choice of a message) to the others' strategies $\sigma^{*N\setminus\{j\}}$, that is,

$$\forall\, m^j \in M^j : Eu^j(g(\sigma^*) \mid t^j) \geq Eu^j(g(m^j, \sigma^{*N\setminus\{j\}}) \mid t^j).$$

Now, the *revelation principle* guarantees that, in designing an optimal mechanism, the principal can restrict the set of admissible mechanisms only to those mechanisms from T to Z (that is, each agent j's message space is his type space T^j) that satisfy Bayesian incentive compatibility. To be precise, it says: *Let $g : M \to Z$ be a mechanism, and let σ^* be a Bayesian equilibrium of g. Set $z_g^* := g \circ \sigma^*$, the equilibrium random outcome. Then, there exists a mechanism $g' : T \to Z$ such that the honest-strategy bundle (the identity function from T to T) is a Bayesian equilibrium of g' and gives rise to z_g^* as its equilibrium random outcome.*

The proof of this principle is extremely simple: Let σ^* be a Bayesian equilibrium relative to a given mechanism $g : M \to Z$. Then, the mechanism $g \circ \sigma^* : T \to Z$ is the required Bayesian incentive-compatible mechanism.

The essential implication of the revelation principle is the computational convenience for the principal's designing: The principal may assume *without loss of generality* that his available mechanisms are Bayesian incentive-compatible. In the light of this principle, one might conjecture that also in the Bayesian society the players may choose Bayesian incentive-compatible strategies *without loss of generality*. But this is false. We will make this point for the extreme case, $T(\pi^j) = T$ for all $j \in N$ (so that $T(S) = T$ for all $S \in \mathcal{N}$).

For the present Bayesian game $\mathcal{BG}(x^S, \bar{x}^{N\setminus S})$ that follows agreement of a strategy bundle $x^S \in F'^S(\bar{x})$, let σ^{*S} be its Bayesian equilibrium. The problem is that the function $x^S \circ \sigma^{*S} : T \to C^S$, while satisfying the measurability and the Bayesian incentive compatibility conditions, may not be a member of $F^S(\bar{x})$.

As a counterexample to disprove the revelation principle in the Bayesian society, consider example 4.1.4. The initial endowment bundle e is an attainable and private measurable strategy bundle. The Bayesian game $\mathcal{BG}(e)$ that follows agreement of e has the unique Bayesian equilibrium σ^*,

$$\sigma^{*j}(a^j) = \sigma^{*j}(b^j) = b^j.$$

The key issue here is that this Bayesian equilibrium is not feasible, while this issue does not arise in the principal-agent framework. The strategy bundle $e \circ \sigma^*$ is a constant function, $e^j \circ \sigma^* : t^j \mapsto 2$, $j \in N$, so is private measurable and Bayesian incentive compatible, but is not attainable.

4.2 Mediator-Based Approach

Vohra (1999) proposed the *mediator-based approach* within the framework of the Bayesian pure exchange economy $\mathcal{E}_{pe}$. Recall that $T(S)$ is an *a priori* given domain of strategies in coalition S, satisfying $\bigcup_{j \in S} T(\pi^j) \subset T(S) \subset T$. There are two possible scenarios for his approach. The first one is standard, and is described by Vohra (1999, page 124, the second paragraph). He postulates that there is an enforcement agency (mediator) for each coalition S, who enforces a coalitionally agreed upon $\mathcal{T}^S$-measurable strategy bundle, a net trade bundle $\{z^j\}_{j \in S}$, $z^j : T(S) \to \mathbf{R}^l$, such that $\sum_{j \in S} z^j(t^S) = \mathbf{0}$ for every $t \in T(S)$. Notice the dependency of z^j on t^S. When players' types are still private information, each player j communicates his type t^j to the mediator, who, having received the reported type profile t^S, enforces each player j to take the promised action $z^j(t^S)$. The mediator does not know the true type profile, and this fact could create the possibility of j's misrepresenting his type. Suppose $\bar{t}^S$ is the true type profile. Assuming that the others report their true types, j's *interim* expected utility of honest report is

$$Eu^j(z^j + e^j \mid \bar{t}^j)$$
$$:= \sum_{t^{N \setminus \{j\}}} u^j \left(z^j(\bar{t}^j, t^{N \setminus \{j\}}) + e^j(\bar{t}^j), (\bar{t}^j, t^{N \setminus \{j\}}) \right) \pi^j(t^{N \setminus \{j\}} \mid \bar{t}^j).$$

On the other hand, assuming also that the others report their true types, j's *interim* expected utility of dishonest report $\tilde{t}^j$ is

$$Eu^j(z^j(\tilde{t}^j, \cdot) + e^j \mid \bar{t}^j)$$
$$:= \sum_{t^{N \setminus \{j\}}} u^j \left(z^j(\tilde{t}^j, t^{N \setminus \{j\}}) + e^j(\bar{t}^j), (\bar{t}^j, t^{N \setminus \{j\}}) \right) \pi^j(t^{N \setminus \{j\}} \mid \bar{t}^j).$$

Vohra postulates Bayesian incentive compatibility that misrepresentation is not worthwhile. Extending his condition to the Bayesian society is straightforward.

CONDITION 4.2.1 (Vorha, 1999) Coalition S's $\mathcal{T}^S$-measurable strategy bundle z^S in the Bayesian pure exchange economy is Bayesian incentive-compatible, in the sense that

$$\forall j \in S : \forall \bar{t} \in T(S) : \forall \tilde{t} \in \bigcap_{i \in S \setminus \{j\}} \operatorname{supp} \pi^i(\cdot \mid \bar{t}^i) :$$
$$Eu^j(z^j + e^j \mid \bar{t}^j) \geq Eu^j(z^j(\tilde{t}^j, \cdot) + e^j \mid \bar{t}^j).$$

Although Vohra emphasizes importance of the mediator's role, in our view the need for a mediator is the vital weakness of the model in descriptive cooperative theory.

Some theorists have called a model with a mediator the "reduced form." Unlike, e.g., the reduced normal-form game obtained from an extensive game, however, nobody has ever explained how a model without a mediator is reduced to a model with a mediator.

There is an oral tradition among some of the contemporary profession to assert that one can postulate the presence of a mediator *without loss of generality*. In order to examine this assertion, we provide the second scenario for Vohra's mediator-based approach now; it attempts to eliminate the mediator from the scene. We will conclude, however, that this attempt is not completely accomplished, so that the difficulty remains. [The readers who are not interested in this oral tradition can skip the following discussion through the end of example 4.2.2.] *Stage 1.* The members of coalition S agree on a strategy bundle $\{z^j\}_{j \in S}$, assuming that their private information will have been fully pooled by the time of strategy execution, that is, they design each z^j so that it is $\mathcal{T}^S$-measurable. Every member j knows the functional form of his colleague i's strategy z^i. *Stage 2.* At the time the members' true types $\bar{t}^j$, $j \in S$, are private information, they simultaneously and independently communicate their types each other. Assuming honest communication of his colleagues, member j's best action is honest communication of his true type. The true type profile $\bar{t}^S$ is thus transmitted to every member. *Stage 3.* Each member j takes the promised action $z^j(\bar{t}^S)$.

There remains one uneasiness about the above scenario: In stage 2, player j evaluates his possible communication based upon the *interim* probability (the conditional probability given $\bar{t}^j$). At this time, he has not made his choice of net trade yet. In stage 3, however, when he is about to make his choice, he can evaluate his choice based upon the *ex post* probability (the conditional probability given $\bar{t}^S$),[3] and according to this updated probability his decision in stage 2 may not have been optimal. In case his decision in stage 2 turns out to be suboptimal, he may refuse to act as promised in stage 3. This point is illustrated in the following simplest example.

EXAMPLE 4.2.2 Consider the Bayesian pure exchange economy with one commodity ($l = 1$) and two consumers ($N = \{1, 2\}$), in which each consumer's type space has two elements ($T^j = \{t_1^j, t_2^j\}$), all type profiles have the equal *ex ante* probability ($\pi(t) = 1/4$), his utility function depends linearly only on his consumption ($u^j(c^j, t) = c^j$), and his endowment is constant ($e^j(t_1^j) = e^j(t_2^j) = 1$). In this case, $T(S) = T$ for all $S \in \mathcal{N}$.

---

[3]This is the *ex post* probability, indeed, since his net trade plan z^j is $\mathcal{T}^S$-measurable.

4: Bayesian Incentive Compatibility Requirement

Define the net trade plan bundle z for the grand coalition by

$$z^1(t) := \begin{cases} -1 & \text{if} \quad t = (t_1^1, t_1^2) \\ 1 & \text{if} \quad t = (t_1^1, t_2^2) \\ 1 & \text{if} \quad t = (t_2^1, t_1^2) \\ -1 & \text{if} \quad t = (t_2^1, t_2^2), \end{cases}$$
$$z^2(t) := -z^1(t).$$

This plan satisfies attainability and Bayesian incentive compatibility (condition 4.2.1). Let $\bar{t}$ be any true type profile, say $\bar{t} = (t_1^1, t_1^2)$. Then consumer 1 ends up with the final consumption of 0 at stage 3, which is less than his initial endowment at this state. Consumer 1 will break off from the grand coalition, taking back his initial endowment. □

The mediator-based approach without a mediator thus postulates a corporate (coalitional) atmosphere which forces its members to always act according to an agreed upon strategy bundle. It is this invisible enforcement atmosphere that we label as the "mediator." In reality, however, the effectiveness of this kind of mediator is questionable.

The private information case, together with the associated private measurability condition, postulates the safe attitude of each coalition that it avoids to design those mechanisms that could make its member reluctant to act at the time of contract execution. Of course, some members of a coalition may *ex post* regret the actions they have made at the *interim* period. But it is too late; what's done is done. This theory successfully explains players' actions.

While the Bayesian incentive compatibility for the private information case (presented in section 4.1) or for the information-revelation case (via contract execution – to be presented in section 9.1) reflects the information system available at the time of making a choice (action), the Bayesian incentive compatibility for the mediator-based cases (presented in sections 4.2, 4.3 and 9.3) does not.

The Bayesian strategic cooperative game theory, pioneered by Wilson (1978), and subsequently developed by Yannelis (1991) as he introduced private measurability, and by Ichiishi and Idzik (1996) as they introduced the general framework and Bayesian incentive compatibility, does not rely on the existence of a mediator. It serves as a theoretical foundation of analyses of the present-day economy, since no mediator, visible or invisible, plays any role in operating the organizations (corporations) in real life. The recent mediator-based approaches left open the question of how to eliminate a mediator.

4.3 Communication Plan as a Part of a Strategy

Bayesian incentive compatibility takes quite a different form when players' strategies involve more than type-profile dependent choices. In an attempt to determine Wilson's communication system (definition 2.1.4) endogenously in an equilibrium (core strategy) of the Bayesian pure exchange economy (example 2.2.1), Yazar (2001) defined player j's strategy in coalition S as a pair of a net trade plan $z^j : T(S) \to \mathbf{R}^l$ and an information substructure $\mathcal{C}^j$. An algebra $\mathcal{C}^j$ on T is called j's *communication plan*, if it is coarser than his private information structure T^j; it is an information structure that j provides to his colleagues as a part of his strategy. When every player i in coalition S chooses strategy $(z^i, \mathcal{C}^i)$, the communication system $\{\mathcal{A}^j\}_{j \in S}$ for S is determined as $\mathcal{A}^j := \mathcal{T}^j \bigvee (\bigvee_{i \in S} \mathcal{C}^i)$. It will turn out that this new definition of strategy $(z^j, \mathcal{C}^j)$ is implicit in the blocking concept used in defining Wilson's fine core (definition 5.1.4).

Yazar's scenario goes as follows. Suppose that coalition S is formed and the members choose a strategy bundle $\{z^j, \mathcal{C}^j\}_{j \in S}$. This means in particular that the members of the coalition communicate the information conveyed by $\mathcal{C}^j$ among themselves. Thus, everybody in the coalition has at least the pooled information structure $\bigvee_{i \in S} \mathcal{C}^i$, so each net trade plan z^j can be made $\bigvee_{i \in S} \mathcal{C}^i$-measurable.

Let $\bar{t} := \{\bar{t}^j\}_{j \in S}$ be the true type profile of coalition S. For simplicity of analysis, we assume that $T(\pi^j) = T(S)$ for every $j \in S$. At the beginning of the *interim* period, everybody has only his private information structure, so member j knows that the event $E := \{\bar{t}^j\} \times T^{N \setminus \{j\}}$ has realized; at this moment, any state $t \in E \bigcap T(S)$ could have occurred from j's point of view. Then, everybody passes on information to his colleagues according to the promised communication plan. Let $C^i(t^i) \times T^{N \setminus \{i\}}$ be the minimal element of $\mathcal{C}^i$ that contains t. Member j thinks that if $t \in E$ occurs and if everybody sends the true information, then everybody receives the additional pooled information that the event $\prod_{i \in S} C^i(t^i)$ has realized. Notice that in the light of the measurability requirement, function z^j is constant on $\prod_{i \in S} C^i(t^i) \bigcap T(S)$, so he can choose net trade $z^j(t)$ no matter which state in $\prod_{i \in S} C^i(t^i) \bigcap T(S)$ is true. Consumer j's *interim* expected utility will then be given as

$$Eu^j(z^j(t) + e^j(\bar{t}^j) \mid \bar{t}^j)$$
$$:= \sum_{t'^{N \setminus \{j\}} \in T^{N \setminus \{j\}}} u^j\left(z^j(t) + e^j(\bar{t}^j), (\bar{t}^j, t'^{N \setminus \{j\}})\right) \pi(t'^{N \setminus \{j\}} \mid \bar{t}^j).$$

Member j can pass on to his colleagues false information $C'^j \times T^{N \setminus \{j\}} \in$

4: Bayesian Incentive Compatibility Requirement

C^j. Then, he thinks that if $t \in E$ occurs and if everybody else passes on to the others the true information according to the promised communication plan, the additional pooled information is that the event $E' := C'^j \times \prod_{i \in S \setminus \{j\}} C^i(t^i)$ has realized. Function z^j is constant on $E' \cap T(S)$. Member j's *interim* expected utility will be $Eu^j(z^j(E' \cap T(S)) + e^j(\bar{t}^j) \mid \bar{t}^j)$.

Yazar's condition of Bayesian incentive compatibility says that no member of a coalition can benefit from providing false information to the other members. Extending her condition to the Bayesian society is straightforward.

CONDITION 4.3.1 (Yazar, 2001) Coalition S's strategy bundle $\{z^j, C^j\}_{j \in S}$ in the Bayesian pure exchange economy is Bayesian incentive-compatible, in the sense that

$$\forall j \in S : \forall \bar{t} \in T(S) : \forall C' \in C^j : \exists t \in \{\bar{t}^j\} \times T^{N \setminus \{j\}} \cap T(S) :$$

$$Eu^j(z^j(t) + e^j(\bar{t}^j) \mid \bar{t}^j) \geq Eu^j(z^j(E' \cap T(S)) + e^j(\bar{t}^j) \mid \bar{t}^j),$$

where $E' := C' \times \prod_{i \in S \setminus \{j\}} C^i(t^i)$.

We remark that Yazar's model without a mediator has the same difficulty as Vohra's mediator-based approach without a mediator, that is, having collected the others' private information, some players may not want to act according to an agreed upon strategy bundle.

We will point out two facts and argue that Vohra's (1999) work is a special case of Yazar's (2001). For notational simplicity, we will assume that $T(S) = T$ for all $S \in \mathcal{N}$, but this assumption may be dropped. Yazar's Bayesian incentive compatibility condition on a strategy bundle with the full communication plan $\{z^j, T^j\}_{j \in S}$ becomes:

$$\forall j \in S : \forall \bar{t}^j \in T^j : \forall \tilde{t}^j \in T^j : \exists t^{N \setminus \{j\}} \in T^{N \setminus \{j\}} :$$

$$\sum_{t'^{N \setminus \{j\}}} u^j \left(z^j(\tilde{t}^j, t^{N \setminus \{j\}}) + e^j(\bar{t}^j), (\bar{t}^j, t'^{N \setminus \{j\}}) \right) \pi(t'^{N \setminus \{j\}} \mid \bar{t}^j)$$

$$\geq \sum_{t'^{N \setminus \{j\}}} u^j \left(z^j(\tilde{t}^j, t^{N \setminus \{j\}}) + e^j(\bar{t}^j), (\bar{t}^j, t'^{N \setminus \{j\}}) \right) \pi(t'^{N \setminus \{j\}} \mid \bar{t}^j).$$

First, she made explicit the measurability of a strategy bundle (condition 3.1.1) with respect to the communication system that is endogenously determined by a communication plan. According to Vohra's scenario, the mediator provides the reported information t^S to each player j (when he tells j to make choice $z^j(t^S)$). This scenario can be viewed as follows: The players have chosen the full communication plan when deciding on the net trade plan z^j. Thus, we can place Vohra's model in Yazar's framework, by viewing Vohra's definition of strategy as a pair (z^j, T^j) of a net trade plan and the full communication plan. Yazar's condition 4.3.1 on a strategy bundle with the full communication plan $\{z^j, T^j\}_{j \in S}$ and Vohra's condition 4.2.1 on a T^S-measurable strategy bundle z^S are then the same in sprit. It is true that there is a difference between the two conditions: According to Yazar's condition, j's *interim* conditional expected utility is computed for *each* possible action he may make, $z^j(\tilde{t}^j, t^{N\setminus\{j\}})$, $t^{N\setminus\{j\}} \in T^{N\setminus\{j\}}$, while according to Vohra's condition, j's *interim* conditional expected utility is obtained by integration with respect to his possible actions; in short, there is separate treatment of $t^{N\setminus\{j\}}$ and $t'^{N\setminus\{j\}}$ in Yazar's condition. However, we view that this difference is minor.

Second, in Yazar's framework an arbitrary communication plan (rather than the full communication plan) is possible as a part of a strategy.

Part II

SOLUTIONS, INFORMATION REVELATION

Chapter 5

Descriptive Solution Concepts

Having presented formal models in part I, we are ready to further specify how players interact within the framework of a given model, and present the associated descriptive solution of the game. From the viewpoint of the principal-agent theory we may say, using the terminology of the principal-agent theory, that most of the literature in the Bayesian cooperative theory to date has dealt with the interactive mode in which each player plays both the role of principal and the role of agent: Players get together to make coordinated strategy choice as principals. After the grand coalition decides on its self-sustaining strategy bundle (descriptive solution of the game), each player execute his agreed strategy as an agent in an *interim* period. The solution is called *ex ante* (*interim*, resp.), if it is agreed upon in the *ex ante* period (in an *interim* period, resp.).

A strategy bundle as a function from the type profile space (message-profile space) to the choice-bundle space may be considered a mechanism. While the principal-agent theory of mechanism design explains a mechanism simply as an optimal solution to the principal's problem, the cooperative theory explains it as an endogenous solution to the game.

5.1 *Interim* Solution Concepts

Wilson (1978) paid attention to availability of communication systems (definition 2.1.4) in defining two notions of core within the framework of Bayesian pure exchange economy (example 2.2.1). It was only long after publication of Wilson (1978) that subsequent authors started addressing the needs for the measurability condition (condition 3.1.1) and the Bayesian incentive compatibility condition (condition 4.1.1, 4.2.1 or 4.3.1). In this section we will present these ideas within the general framework of the Bayesian society (definition 2.1.3).

Let

$$\mathcal{S} := \left(\{C^j, T^j, u^j, \{\pi^j(\cdot \mid t^j)\}_{t^j \in T^j}\}_{j \in N}, \{\mathbf{C}_0^S, T(S), F^S\}_{S \in \mathcal{N}} \right)$$

be a Bayesian society. Define for each prevailing strategy bundle $\bar{x} \in X$ and each coalition S the set of all strategies "feasible" on event $E \subset T(S)$:

$$F_E^S(\bar{x}) := \{x^S|_E : E \to C^S \mid x^S \in F^S(\bar{x})\}.$$

Let $\{\mathcal{A}^j\}_{j \in S}$ be a communication system. Player j's conditional expected utility function of strategy bundle $x \in X$ given $\mathcal{A}^j$ associates with each state $t \in T$ the conditional expected utility of x given the minimal element of $\mathcal{A}^j$ that contains t:

$$\begin{aligned} Eu^j(x \mid \mathcal{A}^j)(t) &:= Eu^j(x \mid A(t)) \\ &= \sum_{s \in T} u^j(x(s), s) \pi^j(s \mid A(t)) \end{aligned}$$

where $A(t)$ is the minimal element of $\mathcal{A}^j$ that contains t, and $\pi^j(\cdot \mid A(t))$ is j's conditional probability on T given event $A(t)$ (so, for example, $Eu^j(x \mid \mathcal{T}^j)(t) = Eu^j(x \mid t^j)$).

The *interim* solutions of $\mathcal{S}$ presented here are broadly classified into two classes, reflecting Wilson's (1978) two core notions for the Bayesian pure exchange economy. They are defined for the general situation in which each consumer has *interim* probabilities $\{\pi^j(\cdot \mid t^j)\}_{t^j \in T^j}$, subjective or objective. The definitions allow for situations in which these *interim* probabilities may not be derived from one *ex ante* probability via the Bayes rule.

The first class is for the situation in which the members of a coalition can use only the null communication system. For an algebra $\mathcal{A}$ on T, define the algebra on $T(S)$ induced by $\mathcal{A}$,

$$\mathcal{A} \bigcap T(S) := \{E \bigcap T(S) \mid E \in \mathcal{A}\}.$$

DEFINITION 5.1.1 (Wilson, 1978) Let $\mathcal{S}$ be a Bayesian society. A strategy bundle x^* is called a *coarse strong equilibrium* of $\mathcal{S}$, if
(i) $x^* \in F^N(x^*)$; and
(ii) it is not true that

$$\exists\, S \in \mathcal{N} : \left(\exists\, E \in \bigwedge_{j \in S} \left(\mathcal{T}^j \bigcap T(S)\right) : E \neq \emptyset \right) : \exists\, x^S \in F_E^S(x^*) :$$

$$\forall\, j \in S : \forall\, t \in E : \quad Eu^j(x^S, x^{*N \setminus S} \mid T^j)(t) > Eu^j(x^* \mid T^j)(t).$$

5: Descriptive Solution Concepts

Condition (i) in definition 5.1.1 is feasibility[1] of strategy bundle x^*; the grand coalition is indeed formed in equilibrium and the members jointly choose this bundle. Condition (ii) is the coalitional stability condition, sometimes called the *group incentive compatibility*; it makes precise the idea that no coalition can improve upon x^* in the following sense: No coalition S can have an event E that all members of S can discern ($E \in \bigwedge_{j \in S} T^j \cap T(S)$) and a feasible strategy x^S on E, such that every member j is made better off with x^S than with x^{*S} at each state in E according to his own private information, passively expecting that the outsiders $N \setminus S$ keep taking the strategy bundle $x^{*N \setminus S}$.

When the outsiders $N \setminus S$ do not influence insider j of S through the feasible-strategy correspondence F^S or through the utility function u^j, for all $S \in \mathcal{N}$ and $j \in S$, that is, when F^S is a constant correspondence and u^j depends only on (c^j, t), the set of the coarse strong equilibria is called the *coarse core*. Indeed, a commodity allocation plan $x^* : T(N) \to \mathbf{R}^{l \cdot \#N}$ in a Bayesian pure exchange economy

$$\mathcal{E}_{pe} := \left\{ X^j, T^j, u^j, e^j, \{\pi^j(\cdot \mid t^j)\}_{t^j \in T^j} \right\}_{j \in N}$$

is said to be in the *coarse core*, if
(i) $x^* \in F^N$; and
(ii) it is not true that

$$\exists S \in \mathcal{N} : \left(\exists E \in \bigwedge_{j \in S} \left(T^j \cap T(S) \right) : E \neq \emptyset \right) : \exists x^S \in F^S_E :$$
$$\forall j \in S : \forall t \in E : \quad Eu^j(x^j \mid T^j)(t) > Eu^j(x^{*j} \mid T^j)(t),$$

where

$$F^S(\bar{x}) := \left\{ x^S : T(S) \to C^S \;\middle|\; \begin{array}{l} x^S \text{ is } T^S\text{-measurable,} \\ \forall\, t : \sum_{j \in S} x^j(t) \leq \sum_{j \in S} e^j(t^j) \end{array} \right\}.$$

In the extreme case of asymmetric information $(\pi^j(\cdot \mid t^j) \gg \mathbf{0}$, so that $T^i(\pi^i) \wedge T^j(\pi^j) = \{\emptyset, T\}$ if $i \neq j$, and consequently $T(S) = T$), the

---

[1] Definitions 5.1.1 and 5.1.4 of coarse core and fine core do not impose the measurability (definition 3.1.1) as a part of the feasibility condition on the solution. It is not clear how far Wilson considered the measurability requirement. In the general formulation of an arbitrarily given state space Ω, in which player j's private information structure is given as an algebra $\mathcal{F}^j$ on Ω, Wilson (1978, page 808, lines 7-10) did require measurability of each strategy with respect to some algebra $\mathcal{F}''$. But the algebra $\mathcal{F}''$ is assumed to be finer than $\bigvee_{j \in N} \mathcal{F}^j$. Then, in the case $\bigvee_{j \in N} \mathcal{F}^j = 2^\Omega$, the algebra $\mathcal{F}''$ is necessarily the finest algebra 2^Ω, so his measurability does not impose any condition on strategies.

coalitional stability condition for the coarse strong equilibrium becomes: the individual rationality condition,

$$\neg \exists j \in N : \exists t^j \in T^j : \exists x^j \in F^j_{\{t^j\} \times T^{N \setminus \{j\}}}(x^*) :$$
$$Eu^j(x^j, x^{*N \setminus \{j\}} \mid t^j) > Eu^j(x^* \mid t^j),$$

and the coalitional stability condition against non-singletons,

$$\neg \, (\exists S : \#S \geq 2) : \exists \, x^S \in F^S(x^*) :$$
$$\forall \, j \in S : \forall \, t^j \in T^j : \quad Eu^j(x^S, x^{*N \setminus S} \mid t^j) > Eu^j(x^* \mid t^j).$$

Here, the sets $\{t^j\} \times T^{N \setminus \{j\}}$, $t^j \in T^j$, are the minimal events that singleton $\{j\}$ can discern, and the entire space T is the only event that all members in a non-singleton S, $\#S \geq 2$, can discern.

The coarse strong equilibrium as an *interim* solution is based on a very conservative attitude towards coalition-formation: Even when player j has the private information $\bar{t}^j$, he wants to make sure before joining a defecting non-singleton coalition S and agreeing on a joint strategy x^S that he is made better off at every type $t^j \in T^j$, including those that he knows have not realized.

If we weaken the individual rationality condition to

$$\neg \, \exists \, j \in N : \exists \, x^j \in F^j(x^*) :$$
$$\forall \, t^j \in T^j : \quad Eu^j(x^j, x^{N \setminus \{j\}} \mid t^j) > Eu^j(x^{*j} \mid t^j),$$

which some of the subsequent literature has done, the resulting weak coalitional stability condition for the coarse core is weaker than the coalitional stability condition for the *ex ante* core; see definitions 5.2.1 and 5.2.2, and remark 5.2.3.

The situation in which players have access only to the null communication system is precisely the private information case. Then we need to impose the private measurability (condition 3.1.1 in which $\mathcal{A}^j = \mathcal{T}^j$ for all $j \in S$) and the Bayesian incentive compatibility (condition 4.1.1) on strategies. Recall the definition of correspondence $\hat{F}^S : X \to C^S$, which takes these two conditions into account (section 4.1).

DEFINITION 5.1.2 Let $\mathcal{S}$ be a Bayesian society, and consider the private information case. A strategy bundle $x^* \in X$ is called a *Bayesian incentive-compatible coarse strong equilibrium* of $\mathcal{S}$, if
(i) $x^* \in \hat{F}^N(x^*)$; and
(ii) it is not true that

$$\exists \, S \in \mathcal{N} : \left(\exists \, E \in \bigwedge_{j \in S} \left(\mathcal{T}^j \bigcap \mathcal{T}(S) \right) : E \neq \emptyset \right) : \exists \, x^S \in \hat{F}^S(x^*) :$$

5: Descriptive Solution Concepts   49

$$\forall\, j \in S : \forall\, t \in E: \quad Eu^j(x^S, x^{*N\setminus S} \mid T^j)(t) > Eu^j(x^* \mid T^j)(t).$$

(Here, we are using $\hat{F}^S(x^*)$ rather than $\hat{F}^S_E(x^*)$ in condition (ii), since it is more appropriate to treat formation of the grand coalition N and formation of a blocking coalition S symmetrically, in that any coalition formation requires planning of choices for *all* contingencies.)

When the correspondence $\hat{F}^S$ is constant and each utility function u^j is a function only of (c^j, t), as in the Bayesian pure exchange economy and in the Bayesian coalition production economy, the set of Bayesian incentive-compatible coarse strong equilibria is called the *Bayesian incentive-compatible coarse core*. We leave it to the reader to specialize definition 5.1.2 to the Bayesian incentive-compatible coarse core for the private information case.

In accordance with the mediator-based approach to the Bayesian pure exchange economy $\mathcal{E}_{pe}$, Vohra (1999) introduced his version of the Bayesian incentive-compatible coarse core. We present its extension to the Bayesian society. Define for $\bar{x} \in X$ and $E \in \mathcal{T}^S \cap T(S)$,

$$F^{ic,S}_E(\bar{x}) := \left\{ x^S \in F^S_E(\bar{x}) \;\middle|\; \begin{array}{l} x^S \text{ is } \mathcal{T}^S\text{-measurable, and} \\ \forall\, j \in S : x^j \text{ is Bayesian incentive-compatible} \\ \text{(condition 4.2.1 in which } E \text{ replaces } T(S)\text{)}. \end{array} \right\}.$$

DEFINITION 5.1.3 (Vohra, 1999) Let S be a Bayesian society, and consider the mediator-based approach. A strategy bundle $x^* : T(N) \to C$ is called a *Bayesian incentive-compatible coarse strong equilibrium* of S, if
(i) $x^* \in F^{ic,N}_{T(N)}(x^*)$; and
(ii) if it is not true that

$$\exists\, S \in \mathcal{N} : \left(\exists\, E \in \bigwedge_{j \in S}\left(\mathcal{T}^j \cap T(S)\right) : E \neq \emptyset \right) : \exists\, x^S \in F^{ic,S}_E(x^*) :$$

$$\forall\, j \in S : \forall\, t \in E: \quad Eu^j(x^S, x^{*N\setminus S} \mid T^j)(t) > Eu^j(x^* \mid T^j)(t).$$

We leave it to the reader to specialize definition 5.1.3 to the *Bayesian incentive-compatible coarse core* for the mediator-based approach.

The second class of solutions is for the situation in which each coalition S is *a priori* endowed with a family of *feasible* communication systems, $C(S)$.

DEFINITION 5.1.4 (Wilson, 1978) Let S be a Bayesian society. For each coalition S, let $C(S)$ be an *a priori* given family of feasible communi-

cation systems which contains the full communication system.[2] A strategy bundle x^* is called a *fine strong equilibrium* of $\mathcal{S}$ with $C(S)$, $S \in \mathcal{N}$, if
(i) $x^* \in F^N(x^*)$; and
(ii) it is not true that

$$\exists\, S \in \mathcal{N} : \exists\, \{\mathcal{A}^j\}_{j \in S} \in C(S) :$$
$$\left(\exists\, E \in \bigwedge_{j \in S} \left(\mathcal{A}^j \bigcap T(S) \right) : E \neq \emptyset \right) : \exists\, x^S \in F_E^S(x^*) :$$
$$\forall\, j \in S : \forall\, t \in E :$$
$$Eu^j(x^S, x^{*N\setminus S} \mid \mathcal{A}^j)(t) > Eu^j(x^* \mid \mathcal{A}^j)(t).$$

When the outsiders $N \setminus S$ do not influence insider j of S through the feasible-strategy correspondence F^S or through the utility function u^j, for all $S \in \mathcal{N}$ and $j \in S$, the set of the fine strong equilibria is called the *fine core*. We leave it to the reader to specialize definition 5.1.4 to the fine core.

We will make three remarks on the fine strong equilibrium concept; the remarks will prompt us to present another *interim* solution concept in the present class.

First, an important question that Wilson left open to future research is clarification of the process according to which the members of coalition S come to be endowed with communication systems $C(S)$. More generally, the process of updating information structure from $\mathcal{T}^j$ to a finer structure needs to be explained. Depending upon specification of such information-revelation process, the solution to the game as given in definition 5.1.4 may no longer be appropriate. This issue will be picked up in chapter 9. In presenting another *interim* solution concept here, therefore, we will confine ourselves only to the private information case (so no communication system other than the null communication system can be used).

Second, there is asymmetry between the grand coalition's formation and a blocking coalition's formation. We have already noted the same asymmetry in discussing the coarse strong equilibrium (in the sentence that immediately follows definition 5.1.2). This asymmetry becomes more prominent in the fine strong equilibrium. Indeed, the coalitional stability condition includes as one of the requirements,

$$\neg\, \exists\, S \in \mathcal{N} : \exists\, t^S \in T^S : \exists\, x^S \in F^S_{\{t^S\} \times T^{N \setminus S}}(x^*) :$$
$$\forall\, j \in S :\ Eu^j(x^S, x^{*N\setminus S} \mid t^S) > Eu^j(x^* \mid t^S),$$

---

[2] Another interpretation of Wilson (1978, p.813, the third paragraph) is that each family $C(S)$ is given as the family of all communication systems for S, $\{\{\mathcal{A}^j\}_{j \in S} \mid \mathcal{T}^j \subset \mathcal{A}^j \subset \mathcal{T}^S\}$.

since the full communication system is available. On the one hand, the grand coalition needs to decide on choice bundles *contingent on all type profiles*. On the other hand, a blocking coalition S needs to decide only on *one* choice bundle $x^S(t^S) \in C^S$ for its formation. Here, we are assuming that x^S does not depend on the outsiders' types, $t^{N \setminus S}$. So, the domain of a function $x^S \in F^S_{\{t^S\} \times T^{N \setminus S}}(x^*)$ is the singleton, $\{t^S\}$, and consequently the space $F^S_{\{t^S\} \times T^{N \setminus S}}$ is identified with a subset of the choice set, C^S. In presenting another *interim* solution concept here, we will remove this asymmetry, by postulating that coalition formation requires planning of choices for all contingencies.

Third, the use of the full communication system implies that a fine strong equilibrium enjoys the strong coalitional stability property that it cannot be improved upon by any coalition *regardless of its type profile*. In presenting another *interim* solution concept here, we will keep this strong property. We thus have:

DEFINITION 5.1.5 Let $\mathcal{S}$ be a Bayesian society, and consider the private information case. A strategy bundle $x^* \in X$ is called an *interim Bayesian incentive-compatible strong equilibrium*, if
(i) $x^* \in \hat{F}^N(x^*)$; and
(ii) it is not true that

$$\exists S \in \mathcal{N}: \exists t^S \in T^S : \exists x^S \in \hat{F}^S(x^*) :$$
$$\forall j \in S: \quad Eu^j(x^S, x^{*N \setminus S} \mid t^j) > Eu^j(x^* \mid t^j).$$

When the outsiders $N \setminus S$ do not influence insider j of S through the feasible-strategy correspondence F^S or through the utility function u^j, for all $S \in \mathcal{N}$ and $j \in S$, the set of *interim* Bayesian incentive-compatible strong equilibria is called the *interim Bayesian incentive-compatible core*. We leave it to the reader to specialize definition 5.1.5 to the *interim* Bayesian incentive-compatible core.

A very specific instance of the *interim* Bayesian incentive-compatible core was used in Ichiishi and Sertel's (1998) study of a profit-center game (example 2.2.3).

We view that the coalitional stability condition in the Bayesian incentive-compatible *interim* strong equilibrium is too strong. This is also suggested by non-existence results in many Bayesian societies satisfying the "standard" assumptions; see, e.g., example 10.1.3. Within the framework of Bayesian pure exchange economy, however, there are clear-cut conditions under which an *interim* Bayesian incentive-compatible core is nonempty (theorem 10.1.4). Perhaps the merit of this concept lies in the fact that it serves as a refinement of any appropriate cooperative solution, as long as we ignore the information-revelation processes. One fundamentally important open question is to propose and study a weaker and appropriate *interim* solution concept.

Our final conceptual issue about the *interim* solution is the definition of *contract*. When each player j knows his true type $\bar{t}^j$, it is questionable whether a contract has to specify all his choices contingent on his types that he knows have not realized; that is, a strategy may not be identified with a contract.

In order to see this point, we review here how the principal-agent theory has addressed this issue. Consider a typical insurance contract theory, in which the insured's probability of causing an accident is his type (his private information), the insurer's type is common knowledge, the insurer is the principal, the insureds are the agents, and the insurer designs a full-coverage insurance policy as a pair $f(t)$ of a premium and a deductible for each possible type t of the insured. Function $f : T \to \mathbf{R}^2$ is a principal's strategy, called a mechanism. The optimal mechanism f^* is the principal's equilibrium strategy. Strategy f^* is not a contract; rather, the image of f^* is interpreted as the set of contracts (insurance policies) he offers to the insureds. Due to Bayesian incentive compatibility, insureds of a specific type t voluntarily choose the contract designed for t, namely $f^*(t)$.

In studying the profit center game with incomplete information (example 2.2.3), Ichiishi and Sertel (1998) interpreted the image $x^*(T)$ of an *interim* core strategy bundle x^* as the set of offered contracts. To appropriately define the *interim* contract concept seems to be determined by specificity of economic contents of the model.

5.2 *Ex Ante* Solution Concepts

We turn to the *ex ante* solution concepts. A strategy is identified with a contract here.

Yannelis (1991) addressed the $\mathcal{T}^j$-measurability in the private information case (condition 3.1.1 for the null communication system) in his core analysis of the Bayesian pure exchange economy $\mathcal{E}_{pe}$ (example 2.2.1). We present a somewhat stronger definition than his original private information core concept (see remark 5.2.3 below). Actually we extend his core concept to the strong equilibrium concept for the Bayesian society (definition 2.1.3). Recall the definition of $F'^S(\bar{x})$ as the set of all private measurable feasible strategy bundles of coalition S.

DEFINITION 5.2.1 (Yannelis, 1991) Let $\mathcal{S}$ be a Bayesian society, and consider the private information case. A strategy bundle x^* is called a *private information strong equilibrium*, if

5: Descriptive Solution Concepts

(i) $x^* \in F'^N(x^*)$: and
(ii) it is not true that

$$\exists\, S \in \mathcal{N} : \exists\, x^S \in F'^S : \forall\, j \in S : Eu^j(x^S, x^{*N\setminus S}) > Eu^j(x^*),$$

where $Eu^j(x)$ is the *ex ante* expected utility of x.

Ichiishi and Idzik (1996) incorporated the Bayesian incentive compatibility condition (condition 4.1.1) in the strong equilibrium analysis:

DEFINITION 5.2.2 (Ichiishi and Idzik, 1996) Let S be a Bayesian society, and consider the private information case. A strategy bundle $x^* \in X$ is called an *ex ante Bayesian incentive-compatible strong equilibrium*, if
(i) $x^* \in \hat{F}^N(x^*)$; and
(ii) it is not true that

$$\exists\, S \in \mathcal{N} : \exists\, x^S \in \hat{F}^S(x^*) : \forall\, j \in S : Eu^j(x^S, x^{*N\setminus S}) > Eu^j(x^*).$$

The existence question on these *ex ante* solutions will be addressed in chapters 8 and 10.

When the outsiders $N \setminus S$ do not influence insider j of S through the feasible-strategy correspondence F^S or through the utility function u^j, for all $S \in \mathcal{N}$ and $j \in S$, that is, when F^S is a constant correspondence and u^j depends only on (c^j, t), the set of the *ex ante* Bayesian incentive-compatible strong equilibria is called the *ex ante Bayesian incentive-compatible core*.

REMARK 5.2.3 In both the original definitions of the *ex ante* private information strong equilibrium and the *ex ante* Bayesian incentive-compatible strong equilibrium, the inequality in (ii) was replaced by:

$$\forall\, t^j \in T^j :\ Eu^j(x^S, x^{*N\setminus S} \mid t^j) \geq Eu^j(x^* \mid t^j),$$

with strict inequality for at least one t^j. While this formulation clarifies the relationship with Wilson's *interim* coarse core concept or the *interim* coarse strong equilibrium concept (see the second paragraph following definition 5.1.1), the present condition (ii) is a stronger coalitional stability condition. The existence proofs of Yannelis (1991) and Ichiishi and Idzik (1996) actually establish the existence of these stronger solutions. □

Einy, Moreno and Shitovitz (2001b) looked at the bargaining set of a Bayesian pure exchange economy with a nonatomic measure space of consumers.

5.3 Other Interactive Modes

Another interactive mode studied is a multi-principal, multi-agent relationship. The player set N is partitioned into the set of principals and the set of agents. The principals play a cooperative game, taking into account the agents' reactions to their coordinated strategy bundle. While there is no general theory of this mode, Ichiishi and Koray (2000) studied a specific model of education, a version of Spence' model. In their model, the first-stage game played by the principals have the same feature as the prisoner's dilemma game, so there exists no cooperative equilibrium.

We point out a central question left open in the area of Bayesian noncooperative game: how to capture and formulate the *interim* market mechanism in the general equilibrium framework.

The question includes a sensible definition of competitive equilibrium. The required notion is expected to share many features with the rational expectations equilibrium, but we believe that in order to fully describe the individual price-taking behavior in the market, an individual demand function needs to be well defined on the entire price simplex $\Delta^{l-1} := \{p \in \mathbf{R}^l_+ \mid \sum_{h=1}^{l} p_h = 1\}$, whereas it is defined only on a negligible subset of Δ^{l-1} in the rational expectations equilibrium framework with a finite type-profile space T. To see this last point on the rational expectations equilibrium, suppose the government announces message function (price function) $\mathbf{p} : T \to \Delta^{l-1}$. If consumer i observes a (possibly disequilibrium) price vector p, he believes that the event $\mathbf{p}^{-1}(p)$ ($\subset T$) occurs. He then conditions his probability on T given $\mathbf{p}^{-1}(p)$, and chooses his demand. But the event $\mathbf{p}^{-1}(p)$ could be empty (or it may have probability 0) unless the observed price p is in the range of function, $\mathbf{p}(T)$. Thus the demand cannot be naturally defined outside the range $\mathbf{p}(T)$. If, for example, T is a finite set, then the range $\mathbf{p}(T)$ is also finite, so the demand function is *un*definable almost everywhere.

Once we accomplish the task of formulating the *interim* market mechanism, we can provide a reliable analysis of the market of lemon, and analyses of simultaneous workings of the market resource allocation mechanism and the non-market resource allocation mechanisms instituted in organizations (firms).

5.4 Coexistence of Coalitions

The feasibility condition (i) in various strong equilibrium concepts (definitions 5.1.1, 5.1.2, 5.1.3, 5.1.4, 5.1.5, 5.2.1, 5.2.2) reflects the scenario that the grand coalition is formed in equilibrium. For applications to economies

5: *Descriptive Solution Concepts* 55

with production, however, we frequently need to explain formation and coexistence of several coalitions (firms). A *coalition structure* is a partition of the player set N; it describes coexisting coalitions. It is easy to extend the Bayesian incentive-compatible strong equilibrium concept, *ex ante* or *interim*, so that a coalition structure is realized in equilibrium (see Ichiishi, 1993a). We will not, therefore, address this issue explicitly in this book.

Chapter 6

Normative Solution Concepts

There are substantial amount of works on normative solution concepts for Bayesian cooperative games; for a survey, see, e.g., the introductory section of Rosenmüller (1992). In this chapter, we present several definitions of specific *interim* normative criterion, Pareto efficiency. From the formal point of view, the normative concept of efficiency is weaker than the descriptive concept of core strategy, as the former is defined by (i) feasibility and (ii) stability only against the grand coalition's improvement.

6.1 *Interim* Efficiency Concepts

Corresponding to the *interim* descriptive solution concepts of the coarse core (definition 5.1.1) and the fine core (definition 5.1.4), Wilson (1978) proposed two *interim* efficiency criteria (definitions 6.1.1 and 6.1.3 below) within the framework of Bayesian pure exchange economy (example 2.2.1) and examined these properties for several numerical examples. These criteria can immediately be extended to the Bayesian society, provided that there be no externality in the sense that for each player j, his utility function u^j depends only on his choice and a type profile (externality is allowed to the extent that the utility function u^j may depend on the others' types $t^{N\setminus\{j\}}$ and the feasible-strategy correspondence F^S may depend on the outsiders' strategy-choice $x^{N\setminus S}$; in particular, the no-externality condition in this chapter is weaker than the no-externality condition assumed in Abreu and Matsushima's lemma 4.1.2 on strict Bayesian incentive compatibility).

Let

$$\mathcal{S} := \Big(\{C^j, T^j, u^j, \{\pi^j(\cdot \mid t^j)\}_{t^j \in T^j} \}_{j \in N}, \ \{\mathbf{C}_0^S, T(S), F^S\}_{S \in \mathcal{N}} \Big)$$

be a Bayesian society (definition 2.1.3), and assume that each u^j is defined on $C^j \times T$ (rather than on $C \times T$).

Wilson's first class of efficiency is for the situation in which each coalition can use only the null communication system. Each coalition is given the freedom to design its own strategy bundle, and we address the issue of efficiency among all possible resulting outcomes.

DEFINITION 6.1.1 (Wilson, 1978) Let $\mathcal{S}$ be a Bayesian society, and assume that each u^j is defined on $C^j \times T$ (rather than on $C \times T$). A strategy bundle x^* of $\mathcal{S}$ is said to be *coarse efficient*, if
(i) $x^* \in F^N(x^*)$; and
(ii) if it is not true that

$$\left(\exists E \in \bigwedge_{j \in N} \left(T^j \bigcap T(N) \right) : E \neq \emptyset \right) : \exists\, x \in F^N(x) :$$
$$\forall\, j \in N : \forall\, t \in E : \quad Eu^j(x^j \mid T^j)(t) > Eu^j(x^{*j} \mid T^j)(t).$$

As the coalitional stability condition for the coarse strong equilibrium, the above coarse efficiency condition (ii) is weak in the extreme case of asymmetric information ($\pi^j(\cdot \mid t^j) \gg 0$, so that $T^i(\pi^i) \wedge T^j(\pi^j) = \{\emptyset, T\}$ if $i \neq j$, and $T(N) = T$ if $\#T \geq 2$). Indeed, in the extreme case of asymmetric information and $\#T \geq 2$, condition (ii) reduces to:

$$\neg\, \exists\, x \in F^N(x) : \forall\, j \in N : \forall\, t \in T : Eu^j(x^j \mid T^j)(t) > Eu^j(x^{*j} \mid T^j)(t).$$

Coarse efficiency in this case is weaker than the *ex ante efficiency*, defined by

$$x^* \in F^N(x^*),$$
$$\neg\, \exists\, x \in F^N(x) : \forall\, j \in N : Eu^j(x^j) > Eu^j(x^{*j}).$$

When each coalition can use only the null communication system, however, the members design a strategy bundle which satisfies two conditions: measurability with respect to the available information structure and Bayesian incentive compatibility. We are thus led to consider the private measurable case. Recall the definition of correspondence $\hat{F}^S : X \to X^S$, which takes the two conditions into account (section 4.1) in the private measurable case. The following is a minor variant of Wilson's coarse efficiency.

6: Efficiency

DEFINITION 6.1.2 Let $\mathcal{S}$ be a Bayesian society, and assume that each u^j is defined on $C^j \times T$ (rather than on $C \times T$). A strategy bundle x^* of $\mathcal{S}$ is said to be *Bayesian incentive-compatible coarse efficient*, if
(i) $x^* \in \hat{F}^N(x^*)$; and
(ii) if it is not true that

$$\left(\exists E \in \bigwedge_{j \in N} \left(T^j \bigcap T(N) \right) : E \neq \emptyset \right) : \exists x \in \hat{F}^N(x) :$$
$$\forall j \in N : \forall t \in E : \quad Eu^j(x^j \mid T^j)(t) > Eu^j(x^{*j} \mid T^j)(t).$$

Due to the imposition of Bayesian incentive compatibility, a Bayesian incentive-compatible coarse efficient strategy bundle is in general suboptimal from the coarse efficiency point of view.

Wilson's second class of efficiency could be for the situation in which each coalition S is *a priori* endowed with a family $C(S)$ of feasible communication systems. Actually, he considered the case in which $C(S)$ consists only of the full communication system.

DEFINITION 6.1.3 (Wilson, 1978) Let $\mathcal{S}$ be a Bayesian society, and assume that each u^j is defined on $C^j \times T$ (rather than on $C \times T$). A strategy bundle x^* of $\mathcal{S}$ is said to be *fine efficient*, if
(i) $x^* \in F^N(x^*)$; and
(ii) if it is not true that

$$\left(\exists E \in T^N \bigcap T(N) : E \neq \emptyset \right) : \exists x \in F^N(x) :$$
$$\forall j \in N : \forall t \in E :$$
$$Eu^j(x^j \mid T^N)(t) > Eu^j(x^{*j} \mid T^N)(t).$$

The fine efficiency reduces to:

$$x^* \in F^N(x^*),$$
$$\neg \exists t \in T(N) : \exists x \in F^N(x) : \forall j \in N : u^j(x^j(t), t) > u^j(x^{*j}(t), t).$$

If the type profiles $T(N)$ are considered the states (specifically, if $\bigvee_{j \in N} \mathcal{F}^j = 2^\Omega$, in the general state-space approach given by $\{(\Omega, \mathcal{F}^j)\}_{j \in N}$ – see section 2.3), and if $F^N(x)$ is the family of all selections of $\mathbf{C}_0^N|_{T(N)}$, this last condition is precisely the *ex post efficiency*. As pointed out in section 5.1, Wilson left open to future research the question of how the players come to be endowed with the full communication system.

Holmström and Myerson (1983) adopted the mediator-based approach: When each player reports his type to the mediator, he bases his decision (about which type to report) on his private information. The mediator can then use the full communication system in the light of the collected information, and he orders each player to take action as specified by the agreed-upon strategy. To avoid misrepresentation of types, the strategy has been made to be Bayesian incentive-compatible.

Formally, define for $\bar{x} \in X$,

$$F^{ic,N}(\bar{x}) := \left\{ x \in F^N(\bar{x}) \;\middle|\; \begin{array}{l} x \text{ is } T^N\text{-measurable, and} \\ \forall\, j \in N : x^j \text{ is Bayesian incentive-} \\ \text{compatible (condition 4.2.1).} \end{array} \right\}.$$

Then, Holmström and Myerson's version of *fine efficient* strategy bundle x^* is defined by

$$x^* \in F^{ic,N}(x^*),$$
$$\neg\, \exists\, t \in T(N) : \exists\, x \in F^{ic,N}(x) : \forall\, j \in N :$$
$$Eu^j(x^j \mid T^j)(t) > Eu^j(x^{*j} \mid T^j)(t).$$

We have already pointed out in section 4.2 that a mediator rarely exists in the real world, so any use of a mediator is the vital weakness of the model.

Among Hahn and Yannelis' (1997) various efficiency criteria for the Bayesian pure exchange economy, their concept of *interim* private efficiency is noteworthy. They considered the private information case, so imposed the condition that each strategy be private measurable. Recall the definition of correspondence $F'^S : X \to X^S$, which associates with each $\bar{x} \in X$ the set of private measurable members of $F^S(\bar{x})$. A strategy bundle x^* in a Bayesian society (with no externality) is called *interim private efficient*, if it satisfies Holmström and Myerson's condition of fine efficiency formulated above, except that correspondence $F^{ic,N}$ be replaced by correspondence F'^N. We further impose Bayesian incentive compatibility, and propose the following *interim* efficiency criterion:

DEFINITION 6.1.4 Let $\mathcal{S}$ be a Bayesian society, and assume that each u^j is defined on $C^j \times T$ (rather than on $C \times T$). A strategy bundle x^* of $\mathcal{S}$ is said to be *Bayesian incentive-compatible interim efficient*, if
(i) $x^* \in \hat{F}^N(x^*)$; and
(ii) if it is not true that

$$\exists\, t \in T(N) : \exists\, x \in \hat{F}^N(x) : \forall\, j \in N :$$
$$Eu^j(x^j \mid T^j)(t) > Eu^j(x^{*j} \mid T^j)(t).$$

Certainly, the reader has noticed the similarities between a Bayesian incentive-compatible coarse strong equilibrium (definition 5.1.2) and a Bayesian incentive-compatible coarse efficient strategy bundle (definition 6.1.2), and between a Bayesian incentive-compatible *interim* strong equilibrium (definition 5.1.5) and a Bayesian incentive-compatible *interim* efficient strategy bundle (definition 6.1.4). The similarities imply that in the free world in which pursuit of self-interest is so strong as to utilize coordinated choice of strategies, as long as the correspondence $\hat{F}^N$ is constant (which is frequently justified because the complementary coalition of N is the empty coalition, so cannot provide externalities to N), the outcome of players' interaction is always efficient. Here, "efficiency" should be more appropriately called "constrained efficiency," since in the present context it is subject to the constraints imposed by private measurability and Bayesian incentive compatibility.

One might wonder, then, what is the point of studying the efficiency concepts in the cooperative game. We believe that they serve as normative criteria for evaluating the outcome of a game played according to a different interactive mode (see, e.g., section 5.3). To the best of our knowledge, however, little research has been done in this direction.

It is widely known that the Bayesian incentive compatibility is a major cause of *ex post* (Pareto-)inefficiency. Ichiishi and Sertel (1998) studied the *interim* Bayesian incentive-compatible core (definition 5.1.5) of the profit center game with incomplete information (example 2.2.3) and showed the following by an example: As we will see exactly in section 9.1, the profit center game has the two-*interim*-period framework, the setup period and the manufacturing period. A full-information-revealing core plan reveals the complete information by the end of the setup period (that is, the manufacturing period becomes the *ex post* period). Players then realize the welfare loss created in the setup period. There are two sources of the loss, the information-revelation process and the Bayesian incentive compatibility. Ichiishi and Sertel (1998) introduced the new scenario that the divisions are free to make a re-contract after the setup period, provided that nobody will receive a lower utility level than was promised in the original contract. Such re-contract can remove the inefficiency caused by the Bayesian incentive compatibility. However, *ex post* inefficiency may still persist, because the divisions have made some of the choices at the time when each had no information about the others' types, that is, the loss caused by the information-revelation process is irreversible.

6.2 Coexistence of Coalitions

The feasibility condition (i) and the non-improvability condition (ii) in various *interim* efficiency concepts (definitions 6.1.1 – 6.1.4) assume that the grand coalition is the most efficient. In reality, however, a nontrivial coalition structure may give rise to a more efficient strategy bundle than the grand coalition; this is particularly so in the presence of decreasing returns with respect to the coalition size beyond certain size. To accommodate this possibility, the definitions of the preceding section need to be modified; but this is easy.

Indeed, let $\mathcal{S}$ be a Bayesian society, such that each u^j is defined on $C^j \times T$ (rather than on $C \times T$), and let $\mathcal{C}_0$ be a family of admissible coalition structures, *a priori* given to the model. A pair of a strategy bundle and a coalition structure, $(x^*, \mathcal{C}^*) \in X \times \mathcal{C}_0$, is called *Bayesian incentive-compatible interim efficient*, if

$$\forall\, S \in \mathcal{C}^* : x^{*S} \in \hat{F}^S(x^*),$$

and

$$\neg\, \exists\, (x, \mathcal{C}) \in X \times \mathcal{C}_0 : \quad \forall\, R \in \mathcal{C} : \quad x^R \in \hat{F}^R(x),$$
$$\forall\, j \in N : \quad Eu^j(x^j \mid \mathcal{T}^j)(t) > Eu^j(x^{*j} \mid \mathcal{T}^j)(t).$$

One can extend the other efficiency concepts to the presence of admissible coalition structures analogously.

Suppose Bayesian society $\mathcal{S}$ is derived from Bayesian coalition production economy $\mathcal{E}_{cp}$ (example 2.2.2). In the presence of a coalition structure, the efficiency concepts (definitions 6.1.1 – 6.1.4) defined relative to the model $\mathcal{S}$ may sometimes fail to capture the intrinsic efficiency of the underlying economy $\mathcal{E}_{cp}$. This is true even for the complete information case.

To see this point, consider, for example, the following coalition production economy $\mathcal{E}_{cp}$ with complete information ($\#T = 1$): There are two consumers ($N = \{1, 2\}$) and three commodities, a, b, and c. The consumers can form singleton organizations $\{1\}$ and $\{2\}$ as production firms, or they can work together in the grand coalition $\{1, 2\}$. Suppose that consumer 1 has his unique resource (commodity b, the component e_b^1 of e^1) which consumer 2 does not have ($e_b^2 = 0$), and that consumer 2 has his unique resources (commodity a, the component e_a^2 of e^2) which consumer 1 does not have ($e_a^1 = 0$). Assume that for efficient use of the technology specific to firm $\{1\}$, consumer 2's unique resources (commodity a) is needed in order to produce output c, and that for efficient use of the technology specific to firm $\{2\}$, consumer 1's unique resources (commodity b) is needed in order to produce output c. Assume further that the larger organization $\{1, 2\}$ has no technology to produce output c due to over-congestion.

To achieve efficiency in this economic model, the coalition structure $\{\{1\},\{2\}\}$ needs to be realized, and the initial endowments need to be re-allocated so that firm $\{1\}$ ($\{2\}$, resp.) holds resource e_a^2 (e_b^1, resp.).

In the associated society $\mathcal{S}$ (see example 2.2.2 for construction), however, singleton $\{1\}$'s feasible-strategy set $F^1(\bar{x})$ reflects availability of only e^1, and not e^2, so cannot make efficient use of its technology. Therefore, the coalition structure $\{\{1\},\{2\}\}$ in society $\mathcal{S}$ cannot yield the efficient production of economy $\mathcal{E}_{cp}$. The coalition structure $\{\{1,2\}\}$ in society $\mathcal{S}$ cannot yield the efficient production either, due to its inefficient technology.

If there is a market for trade of commodities, firms $\{1\}$ and $\{2\}$ can trade their commodities b and a, and utilize the acquired needed resource in production. Efficiency in $\mathcal{E}_{cp}$ is thus achieved. This example illustrates that the perfect neoclassical market mechanism may be more efficient than a cooperative resource allocation mechanism. However, welfare comparison of a market with incomplete information and a cooperative game with incomplete information (the latter is the theme of this book) remains an important uncultivated research area.

Chapter 7

Comparisons of Several Core Concepts

The core solutions are classified according to the time at which strategies are chosen: *ex ante*, *interim*, and *ex post*. The core solutions are subject to the constraints imposed upon strategies: measurability, and Bayesian incentive compatibility. There are interesting works within the framework of Bayesian pure exchange economy (or its variant) which compare the effects on the core solution of different timings, and also the effects on the core solution of different conditions imposed. This chapter reviews such works.

7.1 Fine Core and *Ex Post* Core

We present Einy, Moreno and Shitovitz' (2000a) comparison of the fine core and the *ex post* core. We note that the measurability or the Bayesian incentive compatibility is not imposed on these core concepts. They studied the Bayesian pure exchange economy with a nonatomic space of consumers, an incomplete-information version of Aumann's (1964) seminal model, and established that *a fine core allocation plan is an ex post core allocation plan*. This result is not true for the finite setup. The converse is not true either, even for the nonatomic setup. Compare these negative facts with the positive fact that *under some mild assumptions, the fine efficiency is precisely the ex post efficiency in the finite setup* (the paragraph that follows definition 6.3.1).

We start this section by presenting a counter example of a Bayesian pure exchange economy, to assert that a fine core allocation plan may not be an *ex post* core allocation plan in the finite setup.

EXAMPLE 7.1.1 Consider the Bayesian pure exchange economy,

$$\mathcal{E}_{pe} := \{\mathbf{R}_+^2, T^j, u^j, e^j, \pi\}_{j \in N},$$

in which there are two consumers ($N = \{1,2\}$), two commodities (commodities a and b) and two states, consumer 1 is informed ($T^1 = \{\theta_1, \theta_2\}$) but consumer 2 is uninformed (T^2 is a singleton), so that the type-profile space T is identified with T^1, both consumers have the objective *ex ante* probability π on T given by

$$\pi(\theta_1) = \pi(\theta_2) = \frac{1}{2},$$

their von Neumann-Morgenstern utility functions are given by

$$u^1(c_a, c_b, t) = \begin{cases} 2c_a + c_b & \text{if } t = \theta_1, \\ c_a + 2c_b & \text{if } t = \theta_2, \end{cases}$$
$$u^2(c_a, c_b, t) = c_a + 2c_b \quad \text{for all } t \in \{\theta_1, \theta_2\},$$

and their initial endowment functions are constant and identical

$$e^1(t) = e^2(t) = \begin{pmatrix} 1 \\ 1 \end{pmatrix} \quad \text{for all } t \in \{\theta_1, \theta_2\}.$$

We will show that the feasible plan (x^{*1}, x^{*2}), given by

$$x^{*1}(\theta_1) := \begin{pmatrix} 2 \\ 0 \end{pmatrix}, \quad x^{*1}(\theta_2) := \begin{pmatrix} 2 \\ 0.9 \end{pmatrix},$$
$$x^{*2}(\theta_1) := \begin{pmatrix} 0 \\ 2 \end{pmatrix}, \quad x^{*2}(\theta_2) := \begin{pmatrix} 0 \\ 1.1 \end{pmatrix},$$

is an *interim* core plan but is not an *ex post* core plan. In the *interim* period, the grand coalition cannot block this, since the number

$$u^1(x^1(\theta_1), \theta_1) + u^1(x^1(\theta_2), \theta_2) + 2Eu^2(x^2(\cdot))$$

achieves the maximum at (x^{*1}, x^{*2}) for all plans (x^1, x^2) that are feasible in N. In the *interim* period, singleton $\{1\}$ cannot block this, since

$$u^1(x^{*1}(t), t)) > 3 = u^1(e^1(t), t)), \quad t = \theta_1, \theta_2,$$

and singleton $\{2\}$ cannot block this either, since

$$Eu^2(x^{*2}(\cdot)) = 3.1 > 3 = Eu^2(e^2(\cdot)).$$

Therefore, plan (x^{*1}, x^{*2}) is an *interim* core plan. This is not an *ex post* core plan, however, since singleton $\{2\}$ blocks this at $t = \theta_2$. □

7: Comparisons of Core Concepts

We turn to the infinite setup. Einy, Moreno and Shitovitz assume that while there are infinitely many consumers, there are only finitely many information structures, formulated here using finitely many type spaces, T^j, $j \in N$. (The set N is no longer a consumer set; it is an index set for different information structures.) Denote by T the type-profile space, $\prod_{j \in N} T^j$. A type profile may synonymously be called a state.

Let $(A, \mathcal{A}, \nu)$ be a positive measure space of consumers. There are l commodities, indexed by $h = 1, \ldots, l$. Each consumer $a \in A$ has the consumption set $\mathbf{R}_+^l$, his preference relation is represented by a type-profile-dependent von Neumann-Morgenstern utility function, $u(a, \cdot, \cdot) : \mathbf{R}_+^l \times T \to \mathbf{R}$, and his initial endowment at type-profile t is a commodity bundle $e(a,t) \in \mathbf{R}_+^l$. For each t, the function $e(\cdot, t)$ is assumed to be ν-integrable on A. Let $j(a)$ be the index for consumer a's private information structure; his type space is then $T^{j(a)}$. We assume that $j : A \to N$ is $\mathcal{A}$-measurable.

Each consumer $a \in A$ has his subjective *ex ante* probability π^a on T, which is merely assumed to be nonnegative, $\pi^a \geq \mathbf{0}$. (Notice that while there may be infinitely different probabilities, π^a, $a \in A$, there are only finitely many different supports, due to the finiteness of T, so the equivalence of the two approaches discussed in section 2.3 still holds true.) Let supp π^a be the support of π^a, $\{t \in T \mid \pi^a(\{t\}) > 0\}$.

The domain $T(S)$ of strategies of coalition S is *a priori* given so that

$$\bigcup_{a \in S'} \text{supp } \pi^a \subset T(S) \subset T$$

for some $S' \in \mathcal{A}$ for which $S' \subset S$ and $\nu(S \setminus S') = 0$. A feasible strategy bundle for coalition $S \in \mathcal{A}$ is a feasible commodity allocation plan for S, that is, a function $x : S \times T(S) \to \mathbf{R}_+^l$, such that the function $x(\cdot, t)$ is ν-integrable for each state t, and

$$\forall t \in T(S) : \int_S x(a,t)\nu(da) \leq \int_S e(a,t)\nu(da).$$

Notice that Einy, Moreno and Shitovitz do not treat private measurability or Bayesian incentive compatibility.

For each coalition $S \in \mathcal{A}$ and index $i \in N$, define $S^i := \{a \in S \mid j(a) = i\}$, and let $j_+(S)$ be the set of all indeces $i \in N$ such that $\nu(S^i) > 0$. A family of information structures $\{\mathcal{H}^a\}_{a \in S}$ is called a *communication system* for S, if

$$(\forall a \in S : j(a) \in j_+(S)) : T^{j(a)} \subset \mathcal{H}^a \subset \bigvee_{i \in j_+(S)} T^i.$$

It is called *full*, if
$$(\forall\, a \in S : j(a) \in j_+(S)) : \mathcal{H}^a = \bigvee_{i \in j_+(S)} T^i.$$

The following definition (definition 7.1.2) extends Wilson's fine core (definition 5.1.4) to the possibly infinite Bayesian pure exchange economies:

DEFINITION 7.1.2 For each coalition $S \in \mathcal{A}$, let $C(S)$ be an *a priori* given family of communication systems for S which contains the full communication system. A strategy bundle $x^* : A \times T(A) \to \mathbf{R}_+^l$ is said to be in the *fine core*, if
(i) it is attainable in A,
$$\forall\, t \in T(A) : \int_A x^*(a,t)\nu(da) \leq \int_A e(a,t)\nu(da),$$
and
(ii) if it is not true that
$$(\exists\, S \in \mathcal{A} : \nu(S) > 0) : \exists\, \{\mathcal{H}^a\}_{a \in S} \in C(S) :$$
$$\left(\exists\, E \in \bigwedge_{a \in S : j(a) \in j_+(S)} (\mathcal{H}^a \bigcap T(S)) : E \neq \emptyset \right) :$$
$$(\exists\, x : S \times E \to \mathbf{R}_+^l : x(\cdot, t) \text{ is } \nu\text{-integrable}) : \forall\, t \in E :$$
$$\int_S x(a,t)\nu(da) \leq \int_S e(a,t)\nu(da),$$
$$Eu(a, x(a, \cdot) \mid \mathcal{H}^a)(t) > Eu(a, x^*(a, \cdot) \mid \mathcal{H}^a)(t),$$
$$\nu\text{-a.e. in } \{a \in S \mid j(a) \in j_+(S)\}.$$

DEFINITION 7.1.3 A strategy bundle $x^* : A \times T(A) \to \mathbf{R}_+^l$ is said to be in the *ex post core*, if
(i) it is attainable in A,
$$\forall\, t \in T(A) : \int_A x^*(a,t)\nu(da) \leq \int_A e(a,t)\nu(da),$$
and
(ii) if it is not true that
$$(\exists\, S \in \mathcal{A} : \nu(S) > 0) : \exists\, t_0 \in T(S) :$$
$$(\exists\, c : S \to \mathbf{R}_+^l : \nu\text{-integrable}) :$$
$$\int_S c(a)\nu(da) \leq \int_S e(a,t_0)\nu(da),$$
$$u(a, c(a), t_0) > u(a, x^*(a, t_0), t_0), \quad \nu\text{-a.e. in } S.$$

THEOREM 7.1.4 (Einy, Moreno and Shitovitz, 2000a) *Let*

$$\mathcal{E}_{PE} := \Big((A, \mathcal{A}, \nu), \{\mathbf{R}_+^l, T^{j(a)}, u(a, \cdot, \cdot), e(a, \cdot), \pi^a\}_{a \in A} \Big)$$

be a Bayesian pure exchange economy with a nonatomic positive measure space of consumers $(A, \mathcal{A}, \nu)$ and finitely many type spaces $\{T^j\}_{j \in N}$, such that the domains of strategies for coalitions satisfy

$$T(S) \subset T(Q), \text{ for all } Q \in \mathcal{A} \text{ for which } j_+(Q) = N.$$

A fine core allocation plan of $\mathcal{E}_{PE}$ is an ex post core allocation plan of $\mathcal{E}_{PE}$, if
(i) $j_+(A) = N$, *that is, $\nu(A^i) > 0$ for every $i \in N$, where $A^i := \{a \in A \mid j(a) = i\}$;*
(ii) *for every $t \in T$, the map: $A \times \mathbf{R}_+^l \to \mathbf{R}$, $(a, c) \mapsto u(a, c, t)$, is $\mathcal{A} \otimes \mathcal{B}$-measurable, where $\mathcal{B}$ is the Borel σ-algebra of subsets of $\mathbf{R}_+^l$;*
(iii) $\forall t \in T(A) : \int_A e(a, t) \gg 0$;
(iv) *either the function $u(a, \cdot, t) : \mathbf{R}_+^l \to \mathbf{R}$ is continuous and strictly increasing for each $(a, t) \in A \times T$, or it is continuous, increasing, and vanishes on the boundary of $\mathbf{R}_+^l$ for each $(a, t) \in A \times T$.*

Proof Let $x : A \times T(A) \to \mathbf{R}_+^l$ be a strategy bundle, feasible in A. Suppose x is not in the *ex post* core. Then,

$$(\exists\, S \in \mathcal{A} : \nu(S) > 0) : \exists\, t_0 \in T(S) :$$
$$(\exists\, c : S \to \mathbf{R}_+^l : \nu\text{-integrable}) :$$
$$\int_S c(a)\nu(da) \leq \int_S e(a, t_0)\nu(da),$$
$$u(a, c(a), t_0) > u(a, x(a, t_0), t_0), \ \nu\text{-a.e. in } S.$$

Consider the restriction of $\mathcal{E}_{PE}$ to t_0, a nonatomic pure exchange economy with complete information:

$$\mathcal{E}_{PE}(t_0) := \big((A, \mathcal{A}, \nu), \{\mathbf{R}_+^l, u(a, \cdot, t_0), e(a, t_0)\}_{a \in A}\big).$$

Then, the assignment $\hat{x} : A \to \mathbf{R}_+^l$, $a \mapsto x(a, t_0)$, is attainable in A, but is not a core allocation of $\mathcal{E}_{PE}(t_0)$. Therefore, there exists a coalition Q which improves upon $\hat{x}$ in $\mathcal{E}_{PE}(t_0)$, and by Vind (1972) this set Q can be chosen to be large, specifically,

$$\nu(Q) > \nu(A) - \min_{j \in N} \nu(A^j).$$

There exists a ν-integrable function $\hat{x}' : Q \to \mathbf{R}_+^l$ such that

$$\int_Q \hat{x}'(a)\nu(da) \leq \int_Q e(a, t_0)\nu(da),$$
$$u(a, \hat{x}'(a), t_0) > u(a, \hat{x}(a), t_0), \quad \nu\text{-a.e. in } Q.$$

We will show that coalition Q can fine-improve upon x in $\mathcal{E}_{PE}$. In view of the largeness of Q (i.e., $\nu(Q) > \nu(A) - \min_{j \in N} \nu(A^j)$),

$$j_+(Q) = N,$$

so the full communication system of coalition Q provides each member of Q the full information structure $\mathcal{T}^N$. Therefore,

$$\{t_0\} \in \bigwedge_{a \in Q} (\mathcal{T}^N \bigcap \mathcal{T}(Q)).$$

Define $x' : Q \times \{t_0\} \to \mathbf{R}_+^l$ by $x'(a, t_0) := \hat{x}'(a)$. Then,

$$\int_Q x'(a, t_0)\nu(da) \leq \int_Q e(a, t_0)\nu(da),$$

and ν-a.e. in Q,

$$Eu(a, x'(a, \cdot) \mid \mathcal{T}^N)(t_0)$$
$$= u(a, x'(a, t_0), t_0)$$
$$> u(a, x(a, t_0), t_0)$$
$$= Eu(a, x(a, \cdot) \mid \mathcal{T}^N)(t_0).$$

□

It is unlikely that this theorem holds true if we impose the basic requirements (measurability and Bayesian incentive compatibility) on the fine core.

7.2 Private Measurability and Bayesian Incentive Compatibility

Krasa and Shafer (2001) formulated incomplete information differently from the framework of this book (sections 2.1 and 2.3), and defined the private measurability requirement (c.f. chapter 3) and the Bayesian incentive compatibility requirement (c.f. chapter 4) within their own framework. They applied these requirements to the pure exchange economy in order to define the private information core and the Bayesian incentive-compatible core, respectively. Bayesian incentive-compatible strategies do not have to be private measurable in their work, since their purpose is to compare implications of the private measurability condition and of the Bayesian incentive compatibility condition. According to their formulation, the concept of convergence of incomplete information to the complete information is well-defined. Very roughly stated, their results are that *in general a sequence of private information core allocation plans does not converge to any complete information core allocation plan as the incomplete information converges to the complete information*, but that *complete information strict core allocation plans can be the limit of a sequence of incentive-compatible core allocation plans as the incomplete information converges to the complete information*. While there are differences between their formulation and the Bayesian pure exchange economy as defined earlier in this book (definition 2.2.1), their results nevertheless provide insights into the nature of the two requirements.

We present Krasa and Shafer's model. The crucial difference between our model (chapter 2) and Krasa and Shafer's model is that, while the former formulates incomplete information in terms of an information structure (algebra on a state space), the latter formulates incomplete information in terms of a probability on noisy signals (a probability on false revelation of the true state). To re-state the difference: In our model of chapter 2, a consumer can narrow down a set of states which contains the true state, but cannot discern states in that set, and this indiscernability is defined as incompleteness of information. In Krasa and Shafer's model, on the other hand, a consumer is wrongly told with positive probability that an unrealized state were the realized true state, and this falseness is defined as incomplete information. There is, however, a close relationship between the models of chapter 2 and Krasa and Shafer's model; see proposition 7.2.5, corollary 7.2.6 and remark 7.2.7 at the end of this section.

Let N be a finite set of consumers, and let Ω be a finite *state space*. When state $\omega \in \Omega$ is realized, consumer j receives the signal with a positive probability that state $\phi^j \in \Omega$ has occurred (this information may suffer from noise, so that $\phi^j \neq \omega$). Let $\Phi^j := \Omega$, consumer j's signal space. Let $\Phi := \prod_{j \in N} \Phi^j$, the signal bundle space. *Incomplete information* is defined as a probability π on $\Omega \times \Phi$; the number $\pi(\omega, \{\phi^j\}_{j \in N})$ is the *ex ante* objective probability of the situation in which the true state is ω but consumer j receives the signal that ϕ^j occurred, $j \in N$. In the private information case, consumer j observes signal $\phi^j \in \Phi^j$ and infers that event

$E \subset \Omega$ has occurred with probability

$$\frac{\pi(E \times \Omega \times \cdots \times \overbrace{\{\phi^j\}}^{j} \times \cdots \times \Omega)}{\pi(\Omega \times \Omega \times \cdots \times \{\phi^j\} \times \cdots \times \Omega)}.$$

Complete information is defined as a probability $\hat{\pi}$ on $\Omega \times \Phi$ such that $\hat{\pi}(\{(\omega, \omega, \cdots, \omega) \in \Omega \times \Phi \mid \omega \in \Omega\}) = 1$; in other words, for possible realization of any state $\omega \in \Omega$, each consumer later receives the correct information almost surely that ω has occurred. This is re-formulated as: supp $\hat{\pi} \subset \{(\omega, \omega, \cdots, \omega) \in \Omega \times \Phi \mid \omega \in \Omega\}$. A sequence of incomplete information is said to converge to a complete information, if $\pi \to \hat{\pi}$.

The consumption set of each consumer is the nonnegative orthant $\mathbf{R}_+^l$ of the commodity space $\mathbf{R}^l$. Consumer j's preference relation is represented by a state-dependent von Neumann-Morgenstern utility function, $u^j : \mathbf{R}_+^l \times \Omega \to \mathbf{R}$. His initial endowment is a state-dependent commodity bundle, $e^j : \Omega \to \mathbf{R}_+^l$. The *K-S pure exchange economy* is thus given as a list of specified data,

$$\mathcal{E}_{ks}(\pi) := \left(\Omega, \pi, \{\mathbf{R}_+^l, u^j, e^j\}_{j \in N}\right).$$

The *ex ante* period is defined as the period in which nobody knows the true state, but everybody holds probability π, incomplete or complete. The *interim* period is defined as the period in which the true state ω has occurred, but consumer j receives only signal $\phi^j \in \Phi^j$ and initial endowment bundle $e^j(\omega) \in \mathbf{R}_+^l$. The pair $(\phi^j, e^j(\omega))$ is j's *private information*; it may not be correct (that is, it is possible that $\phi^j \neq \omega$).

Consumer j's strategy is a (state, signal bundle)-contingent commodity bundle, $x^j : \Omega \times \Phi \to \mathbf{R}_+^l$. Its *ex ante* expected utility is

$$Eu^j(x^j) := \sum_{(\omega, \phi) \in \Omega \times \Phi} u^j(x^j(\omega, \phi), \omega) \pi(\omega, \phi).$$

All solutions considered here are *ex ante* solutions. Let $\hat{\pi}$ be a complete information probability. A *complete information core allocation plan* of a K-S pure exchange economy $\mathcal{E}_{ks}(\hat{\pi})$ is a strategy bundle $\{x^{*j}\}_{j \in N}$ such that it is feasible in the grand coalition:

$$\sum_{j \in N} x^{*j}(\omega, \omega, \cdots, \omega) = \sum_{j \in N} e^j(\omega), \text{ for all } (\omega, \omega, \cdots, \omega) \in \text{supp } \hat{\pi},$$

and such that it is not improved upon by any coalition:

$$\neg \exists\, S \in \mathcal{N} : \exists\, x^S : \Omega \times \Phi \to \mathbf{R}_+^{l \cdot \#S} :$$

$$\sum_{j \in S} x^j(\omega, \omega, \cdots, \omega) = \sum_{j \in S} e^j(\omega), \text{ for all } (\omega, \omega, \cdots, \omega) \in \text{supp } \hat{\pi},$$

$$\forall\, j \in S : Eu^j(x^j) > Eu^j(x^{*j}),$$

7: Comparisons of Core Concepts

where the expectation is taken with respect to the complete information probability $\hat{\pi}$.

The private information structure $\mathcal{T}^j$ of consumer j is the algebra on $\Omega \times \Phi$, generated by the sets,

$$\{\omega \in \Omega \mid e^j(\omega) = c^j\} \times \Phi, \quad c^j \in \mathbf{R}_+^l,$$

and

$$\Omega \times \Omega \times \overbrace{\cdots \times \{\phi^j\}}^{j} \times \cdots \times \Omega, \quad \phi^j \in \Phi^j.$$

Given a general incomplete information probability π, a *private information core allocation plan* of a K-S pure exchange economy $\mathcal{E}_{ks}(\pi)$ is defined exactly as the above complete information core allocation plan, except that the incomplete information probability π replaces the complete information probability $\hat{\pi}$, and that all strategies, x^{*j} in the grand coalition and x^j in the blocking coalition, are $\mathcal{T}^j$-measurable. Although consumer j cannot observe $(\omega, \phi^{N\setminus\{j\}})$, he can take action $x^j(\omega, \phi)$ in the *interim* period due to the $\mathcal{T}^j$-measurability of x^j.

Consider an alternative scenario in which the members of coalition S can pool their information, even though such pooling may not be justified. Suppose S is to form and to agree on a strategy bundle $\{x^i\}_{i \in S}$ in the *ex ante* period. For feasibility within S, each strategy x^j is $\bigvee_{i \in S} \mathcal{T}^i$-measurable, so that every member can take action $x^j(\omega, \phi)$ by pooling information within S. As consumer j receives information $(\bar{\phi}^j, \bar{c}^j) \in \Phi^j \times \mathbf{R}_+^l$ later in the *interim* period, his conditional expected utility of strategy x^j given $(\bar{\phi}^j, \bar{c}^j)$ is:

$$Eu^j(x^j \mid \bar{\phi}^j, \bar{c}^j)$$
$$:= \sum_{(\omega, \phi^{N\setminus\{j\}}) \in \Omega \times \Phi^{N\setminus\{j\}}} u^j(x^j(\omega, \phi), \omega) \pi(\omega, \phi^{N\setminus\{j\}} \mid \bar{\phi}^j, \bar{c}^j).$$

Consumer j may misrepresent his received signal $\bar{\phi}^j$ as ϕ'^j to his colleagues $S \setminus \{j\}$, and claim the commodity bundle $x^j(\omega, \phi'^j, \phi^{N\setminus\{j\}})$ instead of $x^j(\omega, \bar{\phi}^j, \phi^{N\setminus\{j\}})$, in case $(\omega, \phi^{N\setminus\{j\}})$ is realized. Then, his conditional expected utility of strategy x^j through this wrongdoing given $(\bar{\phi}^j, \bar{c}^j)$ is:

$$Eu^j(x^j(\cdot, \phi'^j, \cdot) \mid \bar{\phi}^j, \bar{c}^j)$$
$$:= \sum_{(\omega, \phi^{N\setminus\{j\}}) \in \Omega \times \Phi^{N\setminus\{j\}}} u^j(x^j(\omega, \phi'^j, \phi^{N\setminus\{j\}}), \omega) \pi(\omega, \phi^{N\setminus\{j\}} \mid \bar{\phi}^j, \bar{c}^j).$$

Consumer j's strategy $x^j : \Omega \times \Phi \to \mathbf{R}_+^l$ is called *Bayesian incentive-compatible*, if misrepresentation of his signal does not increase his *interim* expected utility at the time of taking action:

$$\forall\,(\bar{\phi}^j,\bar{c}^j)\in\Phi^j\times\mathbf{R}_+^l:\forall\,\phi'^j\in\Phi^j:$$
$$Eu^j(x^j\mid\bar{\phi}^j,\bar{c}^j)\geq Eu^j(x^j(\cdot,\phi'^j,\cdot)\mid\bar{\phi}^j,\bar{c}^j).$$

In this definition, we may restrict $\bar{\phi}^j$ to the projection to Ω of the support of π, $\{\omega\in\Omega\mid\exists\,\phi\in\Phi:\pi(\omega,\phi)>0\}$.

Given a general incomplete information probability π, a *Bayesian incentive-compatible core allocation plan* of a K-S pure exchange economy $\mathcal{E}_{ks}(\pi)$ is defined exactly as the above complete information core allocation plan, except that the incomplete information probability π replaces the complete information probability $\hat{\pi}$, and that all strategies in coalition S are $\bigvee_{i\in S}\mathcal{T}^i$-measurable and Bayesian incentive-compatible, for all $S\in\mathcal{N}$.

We turn to presentation of Krasa and Shafer's main results. We have defined the K-S pure exchange economy as a list of specified data, $\mathcal{E}_{ks}(\pi):=\bigl(\Omega,\pi,\{\mathbf{R}_+^l,u^j,e^j\}_{j\in N}\bigr)$. Krasa and Shafer consider a sequence of economies $\{\mathcal{E}_{ks}(\pi^k)\}_{k=1}^{\infty}$, each endowed with incomplete information π^k, and economy $\mathcal{E}_{ks}(\hat{\pi})$ endowed with complete information $\hat{\pi}$, and studied (non-)relationship between the sequence of cores corresponding to $\{\mathcal{E}_{ks}(\pi^k)\}_k$ and the core of $\mathcal{E}_{ks}(\hat{\pi})$, when $\{\pi^k\}_k$ converges to $\hat{\pi}$. Here, the only difference among the members of the sequence and the limit is the probability defining (in)complete information, and the non-probabilistic aspect is the same. We will, therefore, denote the non-probabilistic aspect of the K-S pure exchange economy by $\mathcal{E}_{ks}:=\bigl(\Omega,\{\mathbf{R}_+^l,u^j,e^j\}_{j\in N}\bigr)$.

Krasa and Shafer's first result concerns a generic statement about the non-convergence of a sequence of the private information cores. They consider a *family* of non-probabilistic aspects of K-S pure exchange economies parameterized by $\theta\in\Theta$, $\{\mathcal{E}_{ks,\theta}\}_{\theta\in\Theta}$, in which different parameters specify different preference profiles as follows: The parameter space Θ is identified with an open subset of $\mathbf{R}^{\#\Omega\cdot l\cdot\#N}$. A parameter θ is an $\#N$-tuple $\{\theta^j\}_{j\in N}$ of functions $\theta^j:\Omega\to\mathbf{R}^l$, $j\in N$ (here, each function θ^j is identified with a point in $\mathbf{R}^{\#\Omega\cdot l}$). Each function θ^j determines consumer j's state-dependent von Neumann-Morgenstern utility function, so the latter is written as:

$$u^j(\,\cdot\,;\theta^j):\mathbf{R}^l\times\Omega\to\mathbf{R},\ (c,\omega)\mapsto u^j(c,\omega,\theta^j(\omega)).$$

Thus, given parameter θ and incomplete information probability π, the K-S pure exchange economy is defined as

$$\mathcal{E}_{ks,\theta}(\pi):=\bigl(\Omega,\pi,\{\mathbf{R}_+^l,u^j(\,\cdot\,;\theta^j),e^j\}_{j\in N}\bigr).$$

Krasa and Shafer assume the following mild regularity conditions on the family $\{\mathcal{E}_{ks,\theta}\}_{\theta\in\Theta}$. Conditions (i) and (iii) in assumption 7.2.1 say that

7: Comparisons of Core Concepts

the indifference curves satisfy the standard neoclassical assumptions and that the consumer is risk averse. Condition (ii) says that change in a parameter does influence a utility function. Condition (iv) says that the initial endowment bundle is always strictly positive.

ASSUMPTION 7.2.1 Let θ be any member of Θ. Fix any consumer j, any state ω, and j's any strictly positive consumption bundle $c \gg \mathbf{0}$. Define $\rho := \theta^j(\omega) \in \mathbf{R}^l$.
(i) Each $u^j(c, \omega, \rho)$ is smooth, $D_c u^j(c, \omega, \rho) \gg \mathbf{0}$, and the $l \times l$ matrix $D^2_{cc} u^j(c, \omega, \rho)$ is negative definite;
(ii) The $l \times l$ matrix $D^2_{c\rho} u^j(c, \omega, \rho)$ is non-singular;
(iii) For any sequence $\{c^k\}_k$ in $\mathbf{R}^l$ such that $c^k \gg \mathbf{0}$ for all k, if there exists a commodity h for which $c^k_h \to 0$ as $k \to \infty$, then $\|D_c u^j(c^k, \omega, \rho)\| \to \infty$ as $k \to \infty$;
(iv) $e^j(\omega) \gg \mathbf{0}$.

A property **P** is called a *generic property* of a K-S pure exchange economy, if there exists an open subset $\tilde{\Theta}$ of Θ for which the Lebesgue measure of $\Theta \setminus \tilde{\Theta}$ is zero, such that property **P** holds true in (the non-probabilistic aspect of) economy $\mathcal{E}_{ks,\theta}$ for all $\theta \in \tilde{\Theta}$. In the following theorem, the attainability condition on feasible allocation plans in terms of equality (rather than weak inequality) is crucial.

THEOREM 7.2.2 (Krasa and Shafer, 2001) *Let $\{\mathcal{E}_{ks,\theta}\}_{\theta \in \Theta}$ be a parameterized family of (the non-probabilistic aspects of) K-S pure exchange economies which satisfy assumption 7.2.1. Assume also $l \geq 2$. Then the following is a generic property of a K-S economy:*
Let $\{\pi^k\}_{k=1}^\infty$ be any sequence of incomplete information probabilities, which converges to a complete information probability $\hat{\pi}$, and they satisfy

$$\forall j \in N : \exists \, \omega^j, \omega'^j \in \Omega :$$
$$\omega^j \neq \omega'^j, \; e^j(\omega^j) = e^j(\omega'^j),$$
$$\hat{\pi}(\omega^j, \omega^j, \cdots \omega^j) > 0, \; \hat{\pi}(\omega'^j, \omega'^j, \cdots \omega'^j) > 0,$$
$$\pi^k(\omega^j, \overbrace{\omega^j, \cdots \omega'^j}^{j}, \cdots \omega^j) > 0 \text{ for all } k.$$

Let $x^{(k)}$ be a private information core allocation plan of $\mathcal{E}_{ks,\theta}(\pi^k)$. Then, none of the limit points of the sequence $\{x^{(k)}\}_k$ is a core allocation plan of $\mathcal{E}_{ks,\theta}(\hat{\pi})$.

Krasa and Shafer (2001, p. 457) has an example to show the importance of the assumption $l \geq 2$. Their proof of theorem 7.2.2 is based on the following lemma:

LEMMA 7.2.3 (Krasa and Shafer, 2001) *Let $\{\mathcal{E}_{ks,\theta}\}_{\theta \in \Theta}$ be a parameterized family of (the non-probabilistic aspects of) K-S pure exchange economies which satisfy assumption 7.2.1, and let $\hat{\pi}$ be a complete information probability. Assume also $l \geq 2$. For each consumer j, choose any two distinct states ω^j and ω'^j such that*

$$\hat{\pi}(\omega^j, \omega^j, \cdots, \omega^j) > 0, \ \hat{\pi}(\omega'^j, \omega'^j, \cdots, \omega'^j) > 0.$$

Then the following is a generic property of a K-S economy: There is no ex ante Pareto optimal and individually rational allocation plan $\{x^j\}_{j \in N}$ for which

$$\forall \ j \in N : x^j(\omega^j, \omega^j, \cdots, \omega^j) = x^j(\omega'^j, \omega'^j, \cdots, \omega'^j).$$

Proof of Theorem 7.2.2 Fix any $\theta \in \Theta$. Let ω^j and ω'^j be the two distinct states with the properties assumed in the theorem. It suffices to show that for any private measurable strategy bundle $\{x^{(k),j}\}_{j \in N}$ that is attainable (with equality) in the grand coalition of the economy $\mathcal{E}_{ks,\theta}(\pi^k)$,

$$x^{(k),j}(\omega^j, \omega^j, \cdots, \omega^j) = x^{(k),j}(\omega'^j, \omega'^j, \cdots, \omega'^j);$$

for, if this is the case, any limit point $\{x^{*j}\}_{j \in N}$ of $\{\{x^{(k),j}\}_{j \in N}\}_{k=1}^{\infty}$ also has the property,

$$x^{*j}(\omega^j, \omega^j, \cdots, \omega^j) = x^{*j}(\omega'^j, \omega'^j, \cdots, \omega'^j),$$

so by lemma 7.2.3, in general it cannot be a complete information core allocation plan of $\mathcal{E}_{ks,\theta}(\hat{\pi})$.

In view of $\pi^k \to \hat{\pi}$, we may assume without loss of generality,

$$\forall \ k : \pi^k(\omega^j, \omega^j, \cdots, \omega^j) > 0, \text{ and } \pi^k(\omega'^j, \omega'^j, \cdots, \omega'^j) > 0.$$

Now, since $x^{(k),j}$ is $\mathcal{T}^j$-measurable, the properties of ω^j and ω'^j imply:

$$x^{(k),j}(\omega^j, \overbrace{\omega^j, \cdots, \omega'^j}^{j} \cdots, \omega^j)$$
$$= x^{(k),j}(\omega'^j, \overbrace{\omega'^j, \cdots, \omega'^j}^{j} \cdots, \omega'^j). \quad (7.1)$$

On the other hand, by the $\mathcal{T}^i$-measurability of $x^{(k),i}$,

$$\forall \ i \in N \setminus \{j\} : x^{(k),i}(\omega^j, \overbrace{\omega^j, \cdots, \omega^j}^{j} \cdots, \omega^j) = x^{(k),i}(\omega^j, \overbrace{\omega^j, \cdots, \omega'^j}^{j} \cdots, \omega^j),$$

7: Comparisons of Core Concepts

so that

$$x^{(k),j}(\omega^j, \overbrace{\omega^j, \cdots, \omega'^j}^{j} \cdots, \omega^j) + \sum_{i \in N \setminus \{j\}} x^{(k),i}(\omega^j, \overbrace{\omega^j, \cdots, \omega'^j}^{j} \cdots, \omega^j)$$

$$= \sum_{i \in N} e^i(\omega^j)$$

$$= x^{(k),j}(\omega^j, \omega^j, \cdots, \omega^j \cdots, \omega^j) + \sum_{i \in N \setminus \{j\}} x^{(k),i}(\omega^j, \overbrace{\omega^j, \cdots, \omega^j}^{j} \cdots, \omega^j)$$

$$= x^{(k),j}(\omega^j, \omega^j, \cdots, \omega^j \cdots, \omega^j) + \sum_{i \in N \setminus \{j\}} x^{(k),i}(\omega^j, \overbrace{\omega^j, \cdots, \omega'^j}^{j} \cdots, \omega^j),$$

therefore,

$$x^{(k),j}(\omega^j, \overbrace{\omega^j, \cdots, \omega'^j}^{j} \cdots, \omega^j)$$

$$= x^{(k),j}(\omega^j, \overbrace{\omega^j, \cdots, \omega^j}^{j} \cdots, \omega^j). \tag{7.2}$$

The required result follows from (7.1) and (7.2). □

Proof of Lemma 7.2.3 **Idea of the Proof:** We adopt the following notation in this proof:

$$N = \{1, 2, \ldots, n\},$$
$$\Omega = \{\omega_1, \omega_2, \ldots, \omega_s\},$$
$$U^j(x^j, \theta^j) = \sum_{\omega \in \Omega} \hat{\pi}(\omega, \omega, \ldots, \omega) u^j(x^j(\omega, \omega, \ldots, \omega), \omega, \theta^j(\omega)).$$

The number $U^j(x^j, \theta^j)$ is the *ex ante* expected utility of consumer j's consumption plan x^j in the economy with complete information, $\mathcal{E}_{ks,\theta}(\hat{\pi})$. Also in view of the complete information, we abuse the notation and write $x^j(\omega)$ for $x^j(\omega, \omega, \ldots, \omega)$.

Fix any parameter value $\theta \in \Theta$. An *ex ante* Pareto optimal and individually rational allocation plan $\{x^j\}_{j \in N}$ is necessarily a solution to the following constrained maximization problem: For some $(n-1)$-dimensional (inverse) weight vector $\lambda := (\lambda_2, \lambda_3, \ldots, \lambda_n) \gg \mathbf{0}$,

$$\text{Maximize} \quad U^1(x^1, \theta^1) + \sum_{j=2}^{n} \lambda_j^{-1} U^j(x^j, \theta^j),$$

subject to $\sum_{j=1}^{n} x^j(\omega) = \sum_{j=1}^{n} e^j(\omega),$ for all ω.

Define, therefore, the Lagrangean,

$\mathcal{L}(x, p, \lambda, \theta)$

$:= U^1(x^1, \theta^1) + \sum_{j=2}^{n} \lambda_j^{-1} U^j(x^j, \theta^j) + p \cdot \left(\sum_{j=1}^{n} e^j - \sum_{j=1}^{n} x^j \right),$

where

$x := (x^1, \ldots, x^n) \in \mathbf{R}^{lsn},$
$p := (p_h(\omega))_{1 \leq h \leq l, \omega \in \Omega} \in \mathbf{R}^{ls},$
$\lambda := (\lambda_2, \ldots, \lambda_n) \in \mathbf{R}^{n-1},$
$\theta := (\theta^1, \ldots, \theta^n) \in \mathbf{R}^{lsn}.$

The first-order conditions in vector notation for the maximization problem are:

$$D_{x^1}\mathcal{L} = D_{x^1}U^1(x^1, \theta^1) - p = \mathbf{0}, \tag{7.3}$$
$$\lambda_j D_{x^j}\mathcal{L} = D_{x^j}U^j(x^j, \theta^j) - \lambda_j p = \mathbf{0}, \text{ for } j = 2, \ldots n, \tag{7.4}$$
$$D_p\mathcal{L} = \sum_{j=1}^{n} e^j - \sum_{j=1}^{n} x^j = \mathbf{0}. \tag{7.5}$$

For each $\lambda \gg \mathbf{0}$, the simultaneous equation system (7.3)–(7.5) uniquely determines a Pareto optimal allocation plan x, along with the associated Lagrangean multiplier p. If we substitute another weight vector λ' for λ, the system (7.3)–(7.5) determines another Pareto optimal allocation plan x'. In other words, we can identify each triple (x, p, λ) satisfying (7.3)–(7.5) with a specific Pareto optimal allocation plan. We can thus identify the set of all *ex ante* Pareto optimal, individually rational allocation plans with a subset of

$\{(x, p, \lambda) \mid (x, p, \lambda) \text{ satisfies } (7.3)–(7.5).\}.$

The last line of the statement of lemma 7.2.3 is an additional property, a part of which is re-stated in view of $l \geq 2$,

$$x_1^j(\omega^j) = x_1^j(\omega'^j), \text{ for } j = 1, \ldots, n-1, \tag{7.6}$$
$$x_2^n(\omega^n) = x_2^n(\omega'^n). \tag{7.7}$$

In order to prove the lemma, we are going to show that there exists an open dense subset $\tilde{\Theta}$ of Θ such that for each $\theta \in \tilde{\Theta}$ the equation system (7.3)–(7.7) has no solution (x, p, λ).

Define the function,

$$D(x,p,\lambda,\theta) := \begin{pmatrix} D_{x^1}U^1(x^1,\theta^1) - p \\ D_{x^2}U^2(x^2,\theta^2) - \lambda_2 p \\ \vdots \\ D_{x^n}U^n(x^n,\theta^n) - \lambda_n p \\ \sum_{j=1}^n e^j - \sum_{j=1}^n x^j \\ x_1^1(\omega^1) - x_1^1(\omega'^1) \\ \vdots \\ x_1^{n-1}(\omega^{n-1}) - x_1^{n-1}(\omega'^{n-1}) \\ x_2^n(\omega^n) - x_2^n(\omega'^n) \end{pmatrix}.$$

The simultaneous equation system (7.3)–(7.7) is summarized as:

$$D(x,p,\lambda,\theta) = \mathbf{0}.$$

It suffices to show that there exists an open dense subset $\tilde{\Theta}$ of Θ such that for each $\theta \in \tilde{\Theta}$ the function $D(\cdot,\cdot,\cdot,\theta)$ is transversal to $\{\mathbf{0}\} \subset \mathbf{R}^{ls(n+1)+n}$, since the number of equations is greater than the number of unknowns. By the transversality theorem, it suffices to show that the function D of variables (x,p,λ,θ) is transversal to $\{\mathbf{0}\}$. We will prove this last fact.

Proof of the fact that D is transversal to $\{\mathbf{0}\}$: We take the first-order derivative of function D with respect to (x,p,λ,θ), and obtain the matrix of the coefficients,

$$E := \begin{pmatrix} \tilde{C} & \tilde{B} \\ \tilde{A} & 0 \end{pmatrix}.$$

The elements of the $sl(n+1) \times [sl(n+1) + (n-1)]$ submatrix $\tilde{C}$ are the partial derivatives of the function

$$(x,p,\lambda,\theta) \mapsto \begin{pmatrix} D_{x^1}U^1(x^1,\theta^1) - p \\ D_{x^2}U^2(x^2,\theta^2) - \lambda_2 p \\ \vdots \\ D_{x^n}U^n(x^n,\theta^n) - \lambda_n p \\ \sum_{j=1}^n e^j - \sum_{j=1}^n x^j \end{pmatrix}$$

with respect to (x, p, λ). So after the first $sl(j-1)$ rows, the next block of sl rows is

$$(\mathbf{0}, \cdots, D^2_{x^j x^j} U^j(x^j, \theta^j), \cdots, \mathbf{0}, -\lambda_j I, \cdots, -p, \cdots, \mathbf{0}).$$

The last block of sl rows is

$$(-I, \cdots, -I, \cdots, -I, 0, \cdots, 0, \cdots, 0).$$

The elements of the $sl(n+1) \times sln$ submatrix $\tilde{B}$ are the partial derivatives of the same function with respect to θ. So after the first $sl(j-1)$ rows, the next block of sl rows is

$$(\mathbf{0}, \cdots, D^2_{x^j \theta^j} U^j(x^j, \theta^j), \cdots, \mathbf{0}).$$

All elements of the last sl rows are 0.

The elements of the $n \times [sl(n+1) + (n-1)]$ submatrix $\tilde{A}$ are the partial derivatives of the function

$$(x, p, \lambda, \theta) \mapsto \begin{pmatrix} x_1^1(\omega^1) - x_1^1(\omega'^1) \\ \cdot \\ \cdot \\ \cdot \\ x_1^{n-1}(\omega^{n-1}) - x_1^{n-1}(\omega'^{n-1}) \\ x_2^n(\omega^n) - x_2^n(\omega'^n) \end{pmatrix}$$

with respect to (x, p, λ). The jth row of matrix $\tilde{A}$, $j < n$, is

$$(\overbrace{0, \cdots, 0}^{sl(j-1)}, \underbrace{\overbrace{0, \cdots, 1}^{s}, \cdots, -1}_{\omega^j}, \cdots, 0, \overbrace{0, \cdots, 0}^{s(l-1)}, \overbrace{0, \cdots, 0}^{sl(n-j)}, \overbrace{0, \cdots, 0}^{sl+(n-1)}).$$
$$\underbrace{}_{\omega'^j}$$

The nth row of matrix $\tilde{A}$ is

$$(\overbrace{0, \cdots, 0}^{sl(n-1)}, \overbrace{0, \cdots, 0}^{s}, \underbrace{\overbrace{0, \cdots, 1}^{s}, \cdots, -1}_{\omega^n}, \cdots, 0, \overbrace{0, \cdots, 0}^{s(l-2)}, \overbrace{0, \cdots, 0}^{sl+(n-1)}).$$
$$\underbrace{}_{\omega'^n}$$

To show that D is transversal to $\{\mathbf{0}\}$, we need to prove that the raws of the matrix E are linearly independent. Consider a linear combination of the rows which is equal to $\mathbf{0}$. Let

$$(\overbrace{\cdots}^{sl(j-1)}, \overbrace{w_1^j(\omega_1), \cdots, w_l^j(\omega_s)}^{sl}, \overbrace{\cdots}^{sl(n-j)}, \overbrace{a_1(\omega_1), \cdots, a_n(\omega_s)}^{sn})$$

7: Comparisons of Core Concepts

be the coefficients for the rows of $(\tilde{C}, \tilde{B})$, and let

$$(b_1, \cdots, b_n)$$

be the coefficients for the rows of $(\tilde{A}, \mathbf{0})$. By looking at the columns of

$$\begin{pmatrix} \tilde{B} \\ \mathbf{0} \end{pmatrix},$$

it follows that $w^j D^2_{x^j \theta^j} U^j(x^j, \theta^j) = 0$ for all $j = 1, \ldots n$, and in view of the fact that $D^2_{x^j \theta^j} U^j(x^j, \theta^j)$ is of full rank, we conclude that $w^j = \mathbf{0}$ for all $j = 1, \ldots n$.

For $j < n$, by looking at the column of E corresponding to $x_1^n(\omega^j)$, it follows from $w^n = \mathbf{0}$ that $a_1(\omega^j) = 0$. Similarly, $a_1(\omega'^j) = 0$. Then, by looking at the column of E corresponding to $x_1^j(\omega^j)$, it follows from $w^j = \mathbf{0}$ and $a_1(\omega^j) = 0$ that $b_j = 0$.

Similarly, $b_n = 0$.

It follows from $w = \mathbf{0}$ and $b = \mathbf{0}$ that $a = \mathbf{0}$. □

Krasa and Shafer's second result is a property of one (non-probabilistic aspects of) K-S pure exchange economy $\mathcal{E}_{ks}$, and concerns the convergence of a sequence of the Bayesian incentive-compatible information cores. We present only what we think to be the bottom line of their second result.

A core allocation plan $\{x^j\}_{j \in N}$ of a K-S pure exchange economy with complete information $\mathcal{E}_{ks}(\hat{\pi})$ is called *strict*, if the associated *ex ante* utility allocation cannot be achieved by a proper subcoalition, that is, if

$$\neg \left(\exists S : \emptyset \neq S \subsetneq N \right) : \exists y^S : \Omega \to \mathbf{R}_+^{l \cdot \#S} :$$

$$\sum_{j \in S} y^j(\omega, \omega, \cdots, \omega) \leq \sum_{j \in S} e^j(\omega, \omega, \cdots, \omega), \ \hat{\pi}\text{-a.e.},$$

$$\forall \ j \in S : Eu^j(y^j) = Eu^j(x^j).$$

THEOREM 7.2.4 (Krasa and Shafer, 2001) *Let $\mathcal{E}_{ks}$ be a (non-probabilistic aspects of) K-S pure exchange economy with at least three consumers, for which $u^j(\cdot, \omega)$ is strictly monotone in $\mathbf{R}_+^l$ for all $j \in N$ and all $\omega \in \Omega$.*

Let $\{\pi^k\}_{k=1}^{\infty}$ be any sequence of incomplete information probabilities, which converges to a complete information probability $\hat{\pi}$, such that for all k sufficiently large.

$$\{\omega \in \Omega \mid \exists \ \phi \in \Phi : \pi^k(\omega, \phi) > 0\} \subset \{\omega \in \Omega \mid \hat{\pi}(\omega, \omega, \cdots, \omega) > 0\}.$$

Assume that the economy $\mathcal{E}_{ks}(\hat{\pi})$ possesses strict core allocation plans, and let $x^* \gg \mathbf{0}$ be any such strict core allocation plan. Then, there exists a

$\bigvee_{j \in N} \mathcal{T}^j$-measurable strategy bundle x such that:
(i) *The ex ante utility allocation of x with respect to $\hat{\pi}$ and the ex ante utility allocation of x^* with respect to $\hat{\pi}$ are identical (so x is a core allocation plan of $\mathcal{E}_{ks}(\hat{\pi})$); and*
(ii) *The strategy bundle x is a Bayesian incentive-compatible core allocation plan of $\mathcal{E}_{ks}(\pi^k)$, for all k sufficiently large.*

Proof Let π be any probability on $\Omega \times \Phi$. By abuse of notation, let π be re-defined throughout this proof as the conditional probability of π given $\{e^j\}_{j \in N}$; if π is a complete information probability, the re-defined π is the same as the original π. To be precise, let $\mathcal{P}_e$ be the partition on Ω generated by $\{e^j\}_{j \in N}$, that is, the family of minimal elements of the algebra on Ω generated by the sets,

$$\{\omega \in \Omega \mid e^j(\omega) = c\}, c \in \mathbf{R}_+^l, j \in N,$$

and define

$$\mathcal{P}_e(\omega) := \text{ the member of } \mathcal{P}_e \text{ that contains } \omega,$$

We re-define the numbers,

$$\frac{\pi(\omega,\phi)}{\pi(\Omega \times [\mathcal{P}_e(\omega)]^N)}, \quad \text{if } \pi\left(\Omega \times [\mathcal{P}_e(\omega)]^N\right) > 0 \text{ and } \phi \in [\mathcal{P}_e(\omega)]^N,$$
$$0, \quad \text{otherwise},$$

as $\pi(\omega, \phi)$. Then, the original pooled information structure on $\Omega \times \Phi$, $\bigvee_{i \in N} \mathcal{T}^i$, is precisely the information structure generated by $\{\Omega \times \{\phi\} \mid \phi \in \Phi\}$ on the support of the re-defined π. Therefore, a $\bigvee_{i \in N} \mathcal{T}^i$-measurable strategy x^j can be any function from Φ to $\mathbf{R}^l$.

Let $\hat{\pi}$ be a complete information probability, and let $x^* \gg \mathbf{0}$ be a strict core allocation plan of $\mathcal{E}_{ks}(\hat{\pi})$. Define a strategy bundle $x : \Phi \to \mathbf{R}^{l \cdot \#N}$ by

1. if there exists $\omega \in \Omega$ such that $\phi^i = \omega$ for all i,

$$x^i(\phi) := x^{*i}(\omega, \cdots \omega), \text{ for all } i \in N;$$

2. if there exist $j \in N$ and $\omega \in \Omega$ such that $\phi^j \neq \omega$ and $\phi^i = \omega$ for all $i \in N \setminus \{j\}$,

$$x^i(\phi) := \begin{cases} \mathbf{0} & \text{if } i = j, \\ x^{*i}(\omega, \cdots \omega) + \frac{1}{\#N-1} x^{*j}(\omega, \cdots \omega) & \text{if } i \in N \setminus \{j\}; \end{cases}$$

3. for all other cases,

$$x^i(\phi) := \begin{cases} e^i(\omega) \text{ for all } i \in N & \text{if } \exists\, \omega \in \Omega : \phi \in [\mathcal{P}_e(\omega)]^N, \\ \text{arbitrary} & \text{otherwise.} \end{cases}$$

7: Comparisons of Core Concepts

Then, for any π, complete or incomplete, and for all $(\omega, \phi) \in \text{supp } \pi$ (so that $(\omega, \phi) \in \Omega \times \Phi$ and $\phi \in [\mathcal{P}_e(\omega)]^N$), $\sum_{i \in N} e^i(\phi) = \sum_{i \in N} x^i(\omega)$. (This identity holds true even for the second case, because $\phi^i = \omega \in \mathcal{P}_e(\omega)$ for all $i \in N \setminus \{j\}$ and $\#(N \setminus \{j\}) \geq 2$.)

Clearly, the *ex ante* utility allocation of x with respect to $\hat{\pi}$ is identical to the *ex ante* utility allocation of x^* with respect to $\hat{\pi}$.

For all k sufficiently large, the *ex ante* utility allocation of x with respect to π^k is not *ex ante* improved upon by any proper subcoalition in $\mathcal{E}_{ks}(\pi^k)$, since x^* is a strict core allocation plan. In order to prove the theorem, therefore, it suffices to show that x is Bayesian incentive-compatible with respect to π^k for all k sufficiently large.

Let
$$\hat{\Omega} := \{\omega \in \Omega \mid (\omega, \omega, \cdots, \omega) \in \text{supp } \hat{\pi}\}.$$

By the assumption on the support of π^k, we may assume without loss of generality, $\hat{\Omega} = \{\omega \in \Omega \mid (\omega, \omega, \cdots, \omega) \in \text{supp } \pi^k\}$ for all k.

Choose any $j \in N$ and any k. Consumer j's *interim* expected utility given $\bar{\phi}^j \in \hat{\Omega}$ is

$$\sum_{(\omega, \phi) \in \Omega \times \Phi} u^j(x^j(\phi), \omega) \pi^k(\omega, \phi \mid \bar{\phi}^j)$$

$$= \frac{1}{\pi^k(\Omega \times \{\bar{\phi}^j\} \times \Phi^{N \setminus \{j\}})} \left[u^j(x^j(\bar{\phi}^j, \cdots \bar{\phi}^j), \bar{\phi}^j) \pi^k(\bar{\phi}^j, \bar{\phi}^j, \cdots, \bar{\phi}^j) \right.$$

$$\left. + \sum_{(\omega, \phi^{N \setminus \{j\}}) \neq (\bar{\phi}^j, \bar{\phi}^j, \cdots, \bar{\phi}^j)} u^j(x^j(\bar{\phi}^j, \phi^{N \setminus \{j\}}), \omega) \pi^k(\omega, \bar{\phi}^j, \phi^{N \setminus \{j\}}) \right],$$

but this converges to $u^j\left(x^j(\bar{\phi}^j, \cdots, \bar{\phi}^j), \bar{\phi}^j\right) (= u^j\left(x^{*j}(\bar{\phi}^j, \cdots, \bar{\phi}^j), \bar{\phi}^j\right))$, as $k \to \infty$.

Suppose that consumer j misrepresents $\bar{\phi}^j$ as ϕ'^j ($\neq \bar{\phi}^j$). Then his *interim* expected utility given $\bar{\phi}^j \in \hat{\Omega}$ is

$$\sum_{(\omega, \phi) \in \Omega \times \Phi} u^j(x^j(\phi'^j, \phi^{N \setminus \{j\}}), \omega) \pi^k(\omega, \phi \mid \bar{\phi}^j)$$

$$= \frac{1}{\pi^k(\Omega \times \{\bar{\phi}^j\} \times \Phi^{N \setminus \{j\}})}$$

$$\times \left[u^j(x^j(\overset{j}{\overbrace{\bar{\phi}^j, \cdots \phi'^j}}, \cdots, \bar{\phi}^j), \bar{\phi}^j) \pi^k(\bar{\phi}^j, \bar{\phi}^j, \cdots, \bar{\phi}^j) \right.$$

$$\left. + \sum_{(\omega, \phi^{N \setminus \{j\}}) \neq (\bar{\phi}^j, \bar{\phi}^j, \cdots, \bar{\phi}^j)} u^j(x^j(\bar{\phi}'^j, \phi^{N \setminus \{j\}}), \omega) \pi^k(\omega, \bar{\phi}^j, \phi^{N \setminus \{j\}}) \right],$$

but this converges to $u^j(x^j(\overbrace{\bar{\phi}^j, \cdots \phi'^j}^{j}, \cdots, \bar{\phi}^j), \bar{\phi}^j)$ $(= u^j(\mathbf{0}, \bar{\phi}^j))$, as $k \to \infty$.

In view of strong monotonicity of $u^j(\cdot, \bar{\phi}^j)$ and strict positiveness of x^{*j}, for all k sufficiently large,

$$\sum_{(\omega,\phi) \in \Omega \times \Phi} u^j(x^j(\phi), \omega) \pi^k(\omega, \phi \mid \bar{\phi}^j)$$
$$> \sum_{(\omega,\phi) \in \Omega \times \Phi} u^j(x^j(\phi'^j, \phi^{N \setminus \{j\}}), \omega) \pi^k(\omega, \phi \mid \bar{\phi}^j)$$

Since $\#\Phi^j < \infty$, we conclude that for all k sufficiently large, the above inequality is true for all $\phi'^j \in \Phi^j \setminus \{\bar{\phi}^j\}$, that is, strategy x^j is Bayesian incentive-compatible. □

We close this section by comparing the complete information (the private information structure, resp.) defined in the framework of chapter 2 and the complete information (the private information structure, resp.) defined in the Krasa-Shafer framework. Throughout the rest of this section, we use subscript ks to indicate probabilities on $\Omega \times \Phi$ in the Krasa-Shafer framework (such as π_{ks} and $\hat{\pi}_{ks}$), and reserve the notation without subscript ks for probabilities on the type-profile space T or on the state space Ω in the frameworks of sections 2.1 and 2.3 (such as π and $\hat{\pi}$). This allows us to avoid confusion.

According to the definition of chapter 2, under the complete information, the *ex ante* stage and the *ex post* stage are identical, that is, $\#\Omega = 1$, or almost equivalently (in the case $\#\Omega$ is arbitrary), there exists $\bar{\omega} \in \Omega$ which every consumer knows will occur surely. The latter condition in the framework of chapter 2 is formulated as supp $\hat{\pi} = \{\bar{\omega}\}$. In the Krasa-Shafer framework, the same condition is formulated as supp $\hat{\pi}_{ks} = \{(\bar{\omega}, \bar{\omega}, \cdots, \bar{\omega})\}$. According to Krasa and Shafer's definition, however, complete information is defined merely by supp $\hat{\pi}_{ks} \subset \{(\omega, \omega, \cdots, \omega) \mid \omega \in \Omega\}$.

Let π_{ks} be an incomplete information probability on $\Omega \times \Phi$, as given in the Krasa-Shafer framework. Let e^j be consumer j's initial endowment function. To simplify our discussion, we assume throughout the rest of this section that e^j is a constant function, so that it does not reveal information. (Constancy of e^j is assumed only in order to simplify our argument here, and it can be dispensed with. For another simplification method, see the first paragraph of the proof of theorem 7.2.4.) We have already defined the private information structure in the sense of Krasa-Shafer: an algebra on $\Omega \times \Phi$. In the present case in which the initial endowment functions do not reveal information, it is the algebra generated by the partition, $\{\Omega \times \Omega \times \cdots \times \{\phi^j\} \times \cdots \times \Omega \mid \phi^j \in \Phi^j\}$. This information structure

7: Comparisons of Core Concepts

is different from the information structure defined in chapter 2; the latter is an algebra on the state space Ω. However, we will see in proposition 7.2.5 below that we can recover the information structure of chapter 2 on Ω from the incomplete information probability on $\Omega \times \Phi$. We emphasize that objectiveness of π_{ks} is crucial in our argument here.

From the incomplete information defined in section 2.3 as an algebra on a state space Ω and an *ex ante* objective probability π on Ω, we can construct the associated incomplete information π_{ks} on $\Omega \times \Phi$ in the sense of Krasa-Shafer. Let Ω be a finite state space, and let $\mathcal{F}^j$ be consumer j's private information structure on Ω, $j \in N$. Then, denoting by $\mathcal{P}^j(\omega)$ the minimal element of the algebra $\mathcal{F}^j$ that contains ω, the required probability π_{ks} on $\Omega \times \Phi$ is given by:

$$\pi_{ks}(\omega, \{\phi^j\}_{j \in N}) := \pi(\omega) \times \prod_{j \in N} \pi(\phi^j \mid \mathcal{P}^j(\omega)).$$

Indeed, if state ω occurs, consumer j updates his probability on Ω from π to $\pi(\cdot \mid \mathcal{P}^j(\omega))$. This updated probability is the probability of his signal given occurrence of ω. We are postulating that ϕ^i and ϕ^j are uncorrelated if $i \neq j$. But ω occurs with probability $\pi(\omega)$, hence the above formula for π_{ks}. In the type-profile framework of section 2.1 ($\Omega = T$, $\mathcal{F}^j = \mathcal{T}^j$), the definition reduces to:

$$\pi_{ks}(t, \{\phi^j\}_{j \in N}) := \pi(t) \times \prod_{j \in N} \pi(\phi^j \mid t^j);$$

here the conditional probability $\pi(\cdot \mid t^j)$ is defined on T, rather than on $T^{N \setminus \{j\}}$ – see the first paragraph of section 2.3.

Given the incomplete information $(\Omega, \mathcal{F}^j, \pi)$, $j \in N$, formulated in section 2.3, let $(\Omega \times \Phi, \pi_{ks})$ be the incomplete information derived from it as in the preceding paragraph. Suppose consumer j knows probability π_{ks}, but does not know the underlying information structure $\mathcal{F}^j$ on Ω. The following proposition says that consumer j can deduce $\mathcal{F}^j$ from the knowledge of functional form of π_{ks}. The assumption in the proposition (that is, $\pi \gg 0$) can be made without loss of generality; otherwise, we re-define supp π as Ω. In particular, if we are given the type-profile space (T, π), we set $\Omega := $ supp π. For each possible information that consumer j receives, $\phi^j \in \Phi^j$, define

$$\mathcal{Q}^j(\phi^j) := \{\omega \in \Omega \mid \exists \, \phi^{N \setminus \{j\}} : (\omega, \phi^j, \phi^{N \setminus \{j\}}) \in \text{supp } \pi_{ks}\}.$$

From the knowledge of probability π_{ks}, consumer j knows this set.

PROPOSITION 7.2.5 *Let $\mathcal{E}_{pe}$ be a Bayesian pure exchange economy endowed with a general state space $(\Omega, \{\mathcal{F}^j\}_{j \in N})$, and let $\mathcal{E}_{ks}(\pi_{ks})$ be the*

K-S pure exchange economy constructed from $\mathcal{E}_{pe}$. Assume that $\pi \gg 0$. Consider simultaneous realization of $\bar{\omega} \in \Omega$ and $\bar{\phi}^j \in \Phi^j$, that is, $\bar{\omega}$ is the true state and consumer j observes private information $\bar{\phi}^j$. Then, $\mathcal{Q}^j(\bar{\phi}^j) = \mathcal{P}^j(\bar{\omega})$.

Proof We first claim:
$$\forall\, \omega \in \mathcal{Q}^j(\bar{\phi}^j) : \bar{\phi}^j \in \mathcal{P}^j(\omega).$$
To see this, choose any $\omega \in \mathcal{Q}^j(\bar{\phi}^j)$. Then, $\pi_{ks}(\omega, \bar{\phi}^j, \phi^{N\setminus\{j\}}) > 0$ for some $\phi^{N\setminus\{j\}} \in \Phi^{N\setminus\{j\}}$, so in the light of the formula for π_{ks}, $\pi(\bar{\phi}^j \mid \mathcal{P}^j(\omega)) > 0$, and the claim was proved.

Now, simultaneous realization of $\bar{\omega}$ and $\bar{\phi}^j$ is possible only when $\bar{\omega} \in \mathcal{Q}^j(\bar{\phi}^j)$, so $\bar{\phi}^j \in \mathcal{P}^j(\bar{\omega})$ by the claim. Thus, for all $\omega \in \mathcal{Q}^j(\bar{\phi}^j)$, it follows that $\mathcal{P}^j(\omega) \bigcap \mathcal{P}^j(\bar{\omega}) \ni \bar{\phi}^j$. Since both $\mathcal{P}^j(\omega)$ and $\mathcal{P}^j(\bar{\omega})$ are members of the partition $\mathcal{P}^j$, we have proved
$$\forall\, \omega \in \mathcal{Q}^j(\bar{\phi}^j) : \mathcal{P}^j(\omega) = \mathcal{P}^j(\bar{\omega}).$$
Since $\omega \in \mathcal{P}^j(\omega)$, this establishes $\mathcal{Q}^j(\bar{\phi}^j) \subset \mathcal{P}^j(\bar{\omega})$.

Suppose on the other hand that $\omega \notin \mathcal{Q}^j(\bar{\phi}^j)$. Then, again by the formula for π_{ks}, $\pi(\bar{\phi}^j \mid \mathcal{P}^j(\omega)) = 0$, that is, $\bar{\phi}^j \notin \mathcal{P}^j(\omega)$, so $\mathcal{P}^j(\omega)$ and $\mathcal{P}^j(\bar{\omega})$ are disjoint, and consequently, $\omega \notin \mathcal{P}^j(\bar{\omega})$. $\square$

Clearly, $\bar{\phi}^j \in \mathcal{Q}^j(\bar{\phi}^j)$, so proposition 7.2.5 also says $\mathcal{Q}^j(\bar{\phi}^j) = \mathcal{P}^j(\bar{\phi}^j)$. Thus, $\mathcal{Q}^j := \{\mathcal{Q}^j(\bar{\phi}^j) \mid \bar{\phi}^j \in \Phi^j\}$ is precisely the partition $\mathcal{P}^j$ on Ω, and generates the algebra $\mathcal{F}^j$.

COROLLARY 7.2.6 *Let $\mathcal{E}_{pe}$ be a Bayesian pure exchange economy endowed with a general state space $(\Omega, \{\mathcal{F}^j\}_{j\in N})$, and let $\mathcal{E}_{ks}(\pi_{ks})$ be the K-S pure exchange economy constructed from $\mathcal{E}_{pe}$. Assume that $\pi \gg 0$. Let $x^j : \Omega \times \Phi \to \mathbf{R}_+^l$ be consumer j's strategy in $\mathcal{E}_{ks}(\pi_{ks})$. Then, for every information $\bar{\phi}^j \in \Phi^j$, the interim expected utility given $\bar{\phi}^j$ is:*

$$Eu^j(x^j \mid \Omega \times \Omega \times \cdots \times \{\bar{\phi}^j\} \times \cdots \times \Omega)$$
$$= \sum_{(\omega,\phi^{N\setminus\{j\}}) \in \mathcal{P}^j(\bar{\phi}^j) \times \Phi^{N\setminus\{j\}}} u^j(x^j(\omega, \bar{\phi}^j, \phi^{N\setminus\{j\}}), \omega)$$
$$\times \left[\pi(\omega \mid \mathcal{P}^j(\bar{\phi}^j)) \times \prod_{i\in N\setminus\{j\}} \pi\left(\phi^i \mid \mathcal{P}^i(\phi^i) \bigcap \mathcal{P}^j(\bar{\phi}^j)\right) \right].$$

If x^j is private measurable, then
$$Eu^j(x^j \mid \Omega \times \Omega \times \cdots \times \{\bar{\phi}^j\} \times \cdots \times \Omega)$$
$$= \sum_{\omega \in \mathcal{P}^j(\bar{\phi}^j)} u^j(x^j(\bar{\phi}^j), \omega) \pi(\omega \mid \mathcal{P}^j(\bar{\phi}^j)).$$

Proof By direct substitution of the definition of π_{ks},

$$Eu^j(x^j \mid \Omega \times \Omega \times \cdots \times \{\bar{\phi}^j\} \times \cdots \times \Omega)$$
$$= \sum_{(\omega,\phi) \in \Omega \times \Phi} u^j(x^j(\omega,\phi),\omega)$$
$$\times \left[\pi(\omega \mid \bar{\phi}^j) \times \pi(\phi^j \mid \mathcal{P}^j(\omega), \bar{\phi}^j) \times \prod_{i \in N \setminus \{j\}} \pi(\phi^i \mid \mathcal{P}^i(\omega), \bar{\phi}^j) \right].$$

But by proposition 7.2.5,

$$\pi(\omega \mid \bar{\phi}^j) = \pi(\omega \mid \mathcal{P}^j(\bar{\phi}^j)),$$
$$\pi(\phi^j \mid \mathcal{P}^j(\omega), \bar{\phi}^j) = \begin{cases} 1 & \text{if } \phi^j \in \mathcal{P}^j(\omega) \text{ and } \phi^j = \bar{\phi}^j \\ 0 & \text{otherwise,} \end{cases}$$
$$\pi(\phi^i \mid \mathcal{P}^i(\omega), \bar{\phi}^j) = \pi\left(\phi^i \mid \mathcal{P}^i(\omega) \bigcap \mathcal{P}^j(\bar{\phi}^j)\right) \quad \text{for all } i \neq j,$$

so the required first identity follows.

The second identity follows from the first, if x^j is private measurable, that is, if it is a function only of ϕ^j. □

Notice a striking similarity between the formula for the *interim* expected utility of a private measurable strategy in the Bayesian pure exchange economy (example 2.2.1) and the formula for the *interim* expected utility of a private measurable strategy in the K-S pure exchange economy. However, the two are different: In the former economy, a private measurable strategy from Ω to $\mathbf{R}^l_+$ has to be constant on each minimal element of $\mathcal{F}^j$, while in the latter economy, a private measurable strategy can be any function from $\Phi^j \ (= \Omega)$ to $\mathbf{R}^l_+$. Let $\bar{\phi}^j$ be a signal given to j when $\bar{\omega}$ is the true state. Then, $\mathcal{P}^j(\bar{\phi}^j) = \mathcal{P}^j(\bar{\omega})$, by proposition 7.2.5. In the K-S economy, if the signal contains noise (i.e., if $\bar{\phi}^j \neq \bar{\omega}$), then $x^j(\bar{\phi}^j)$ may be different from $x^j(\bar{\omega})$, so the *interim* expected utility given $\bar{\phi}^j$ may be different from the *interim* expected utility given $\bar{\omega}$. In the Bayesian pure exchange economy (example 2.2.1), on the other hand, the two *interim* expected utilities are the same, because the $\mathcal{F}^j$-measurability stipulates that $x^j(\bar{\phi}^j) = x^j(\bar{\omega})$.

REMARK 7.2.7 We demonstrate by an example that arbitrary information probabilities in Krasa and Shafer's framework cannot always be consistent with the Bayesian pure exchange economy (example 2.2.1), even with a general state space approach of section 2.3 (the approach that is handy in looking at all possible information structures). We also use the same example to verify the content of proposition 7.2.5.

Consider a Bayesian pure exchange economy $\mathcal{E}_{pe}$ with two consumers ($N = \{1,2\}$), endowed with a state space consisting of two states ($\Omega = \{a,b\}$). The objective *ex ante* probability on Ω is given by the probability which assigns density p on $\{a\}$, and $(1-p)$ on $\{b\}$.

There are two possible information structures on Ω: the coarsest algebra $\mathcal{F}_c := \{\emptyset, \{a,b\}\}$, and the finest algebra $\mathcal{F}_f := \{\emptyset, \{a\}, \{b\}, \{a,b\}\}$. Then, there are four possible cases: (case cc) both consumers are endowed with the coarsest information structure $\mathcal{F}_c$; (case cf) consumer 1 is endowed with the coarsest information structure $\mathcal{F}_c$ and consumer 2 is endowed with the finest information structure $\mathcal{F}_f$; (case fc) consumer 1 is endowed with the finest information structure $\mathcal{F}_f$ and consumer 2 is endowed with the finest information structure $\mathcal{F}_c$; (case ff) both consumers are endowed with the finest information structure $\mathcal{F}_f$. The probabilities π_{ks} on $\Omega \times \Phi$ for the four cases are given in the following table.

	case cc	case cf	case fc	case ff
$\pi_{ks}(a,a,a)$	p^3	p^2	p^2	p
$\pi_{ks}(a,a,b)$	$p^2 \cdot (1-p)$	0	$p \cdot (1-p)$	0
$\pi_{ks}(a,b,a)$	$p^2 \cdot (1-p)$	$p \cdot (1-p)$	0	0
$\pi_{ks}(a,b,b)$	$p \cdot (1-p)^2$	0	0	0
$\pi_{ks}(b,a,a)$	$p^2 \cdot (1-p)$	0	0	0
$\pi_{ks}(b,a,b)$	$p \cdot (1-p)^2$	$p \cdot (1-p)$	0	0
$\pi_{ks}(b,b,a)$	$p \cdot (1-p)^2$	0	$p \cdot (1-p)$	0
$\pi_{ks}(b,b,b)$	$(1-p)^3$	$(1-p)^2$	$(1-p)^2$	$(1-p)$

Values of $\pi_{ks}(\omega, \phi^1, \phi^2)$:

Thus, in order to be consistent with the framework of section 2.3, the probability π_{ks} as a point in $\mathbf{R}^8$ should be one of the four column vectors in the above table for some $p \in [0,1]$. For example, probability $(1/9, 3/9, 0, 0, 0, 0, 3/9, 2/9) \in \mathbf{R}^8$ cannot be consistent with the framework of section 2.3.

Assume $0 < p < 1$, and consider case cf. From the probability π_{ks}, consumer 1 knows

$$Q(\phi^1) := \{\omega \in \Omega \mid \exists\, \phi^2 : \pi_{ks}(\omega, \phi^1, \phi^2) > 0\}$$
$$= \{a,b\} \text{ for all } \phi^1 \in \Phi^1,$$

so he deduces that he has the coarsest information structure on Ω. Consumer 2 on the other hand knows

$$Q(\phi^2) := \{\omega \in \Omega \mid \exists\, \phi^1 : \pi_{ks}(\omega, \phi^1, \phi^2) > 0\}$$
$$= \begin{cases} \{a\} & \text{if } \phi^2 = a, \\ \{b\} & \text{if } \phi^2 = b, \end{cases}$$

7: Comparisons of Core Concepts   89

so he deduces that he has the finest information structure on Ω.

Another fact in this example is that if $0 < p < 1$, there exists a complete information probability in Krasa and Shafer's sense iff both consumers have the finest information structure $\mathcal{F}_f$ (case ff), and in this case, the complete information probabilities are the only probabilities consistent with the framework of section 2.3.

In each of the four cases, a sequence of incomplete information probabilities (in the Krasa-Shafer sense) converges to the complete information of sure occurrence of $\{a\}$ (in the sense of chapter 2), as $p \to 1$. □

Chapter 8

Existence

Chapter 5 introduced descriptive solutions to a Bayesian society. This chapter presents general existence results for these concepts. Research on existence of an *interim* solution is still at its infant stage now, while positive results have been established on the existence of an *ex ante* solution. The results to date are heavily based on the static cooperative game theory, so the appendix to this chapter briefly reviews some static cooperative game theory.

8.1 *Interim* Solutions

Wilson (1978) established a coarse core nonemptiness theorem for a Bayesian pure exchange economy, by constructing from the economy a particular non-side-payment game, and by showing that the latter game satisfies the assumptions of Scarf's core nonemptiness theorem (theorem 8.A.2 in the appendix to this chapter). We will pick up his essential idea, and summarize it as proposition 8.1.1. The proposition should not be considered a "result"; rather, it is a possible stepping-stone for a positive result on a Bayesian incentive-compatible coarse strong equilibrium or the Bayesian incentive-compatible coarse core; indeed, it will be used in section 10.1. The reader who does not want to bother with technical details can skip the following paragraphs and go directly to the last paragraph of this section.

Let

$$\mathcal{S} := \left(\{C^j, T^j, u^j, \{\pi^j(\cdot \mid t^j)\}_{t^j \in T^j} \}_{j \in N}, \{\mathbf{C}_0^S, T(S), F^S\}_{S \in \mathcal{N}} \right)$$

be a Bayesian society (definition 2.1.3). We introduce the concept of an *agent* as a pair (j, t^j) of a player and his type, and construct a static society (definition 8.A.3 of the appendix to this chapter) with the agent set $\mathbf{A} := \{(j, t^j) \mid j \in N, t^j \in T^j\}$ as its "player set." Agent (j, t^j)'s strategy space is player j's strategy space X^j in $\mathcal{S}$, and his static utility function is j's conditional expected utility function given t^j, $x \mapsto Eu^j(x \mid t^j)$.

In an agent coalition, any two agents representing the same player, (j, t^j) and (j, t'^j), are postulated to take the same strategy. A strategy bundle in agent coalition $\mathbf{C}$ is then identified with a strategy bundle in the player coalition $S(\mathbf{C}) := \{j \mid (j, t^j) \in \mathbf{C}\}$.

For each $S \in \mathcal{N}$ and each $E \in \bigwedge_{j \in S}(T^j \bigcap T(S))$, define the agent coalition $(S, E) \subset \mathbf{A}$ by

$$(S, E) := \left\{ (j, t^j) \in S \times T^j \mid \emptyset \neq \left(\{t^j\} \times T^{N \setminus \{j\}} \right) \bigcap T(S) \subset E \right\}.$$

For a coalition $\mathbf{C}$ of agents which can be expressed this way, i.e., for

$$\mathbf{C} = (S, E) \text{ for some } (S, E) \in \mathcal{N} \times \bigwedge_{j \in S} \left(T^j \bigcap T(S) \right),$$

the set of feasible strategies when the grand coalition is taking strategy bundle $\bar{x}$ is:

$$\hat{\mathbf{F}}^{\mathbf{C}}(\bar{x}) := \hat{F}^S(\bar{x}).$$

We postulate that the other kind of agent coalitions are not worth forming. To formulate this postulate, we introduce a new (artificial) strategy bundle $\underline{x}$, and define $Eu^j(\underline{x}^j, x^{N \setminus \{j\}} \mid t^j) := -\infty$ for all $x \in X \bigcup \{\underline{x}\}$, and

$$\hat{\mathbf{F}}^{\mathbf{C}}(\bar{x}) := \{\underline{x}^{S(\mathbf{C})}\},$$

if $\mathbf{C} \neq (S, E)$ for any $(S, E) \in \mathcal{N} \times \bigwedge_{j \in S} \left(T^j \bigcap T(S) \right)$.

The set of attainable utility allocations of agent coalition $\mathbf{C}$ given the prevailing strategy bundle $\bar{x}$ is:

$$\hat{V}_{\bar{x}}(\mathbf{C}) := \left\{ u \in \mathbf{R}^{\sum_{j \in N} \#T^j} \mid \begin{array}{l} \exists\, x^{S(\mathbf{C})} \in \hat{\mathbf{F}}^{\mathbf{C}}(\bar{x}) : \forall\, (j, t^j) \in \mathbf{C} : \\ u_{(j, t^j)} \leq Eu^j(x^{S(\mathbf{C})}, \bar{x}^{N \setminus S(\mathbf{C})} \mid t^j) \end{array} \right\}.$$

A family $\mathcal{B}$ of subsets of $\mathbf{A}$ is called *balanced* if there exists $\{\lambda_{\mathbf{C}}\}_{\mathbf{C} \in \mathcal{B}} \subset \mathbf{R}_+$ such that $\sum_{\mathbf{C} \in \mathcal{B}: \mathbf{C} \ni (j, t^j)} \lambda_{\mathbf{C}} = 1$ for every $(j, t^j) \in \mathbf{A}$.

PROPOSITION 8.1.1 *Let $\mathcal{S}$ be a Bayesian society. There exists a Bayesian incentive-compatible coarse strong equilibrium of $\mathcal{S}$, if*
(i) X^j is a nonempty, compact, convex subset of a Hausdorff locally convex topological vector space over $\mathbf{R}$ for every $j \in N$;

(ii) $u^j(\cdot \mid t^j)$ is continuous in X for every $(j, t^j) \in \mathbf{A}$;
(iii) $\hat{F}^S$ is upper and lower semicontinuous in X, and is nonempty- and closed-valued for every $S \in \mathcal{N}$;
(iv) given any $\bar{x} \in X$, and any balanced family $\mathcal{B}$ of subsets of $\mathbf{A}$,

$$\bigcap_{C \in \mathcal{B}} \hat{V}_{\bar{x}}(C) \subset \hat{V}_{\bar{x}}(\mathbf{A}); \quad \text{and}$$

(v) given any $\bar{x} \in X$ and any core utility allocation $\bar{u}$ of game $\hat{V}_{\bar{x}}$, the set

$$\{x^N \in \hat{F}^N(\bar{x}) \mid \forall \, (j, t^j) \in \mathbf{A} : \, u^{(j,t^j)}(x^N) \geq \bar{u}_j\}$$

is convex.

Proof By theorem 8.A.5, there exists a social coalitional equilibrium $x^* \in X$ of the static society with the set $\mathbf{A}$ of agents as its player set. We show that x^* is the required Bayesian incentive-compatible coarse strong equilibrium of $\mathcal{S}$. By the stability condition of a social coalitional equilibrium,

$$\neg \, \exists \, S \in \mathcal{N} : \exists \, E \in \bigwedge_{j \in S} \left(T^j \bigcap T(S) \right) :$$
$$\exists \, x^S \in \hat{F}^S(x^*) : \forall \, (j, t^j) \in (S, E) :$$
$$Eu^j(x^S, x^{*N \setminus S} \mid t^j) > Eu^j(x^* \mid t^j).$$

But this is precisely the coalitional stability condition (ii) for a Bayesian incentive-compatible coarse strong equilibrium. □

No general existence theorem has been established for the *interim* Bayesian incentive-compatible strong equilibrium or for the *interim* Bayesian incentive-compatible core strategy bundle. Nonemptiness of the fine core also remains open. The situation becomes quite different for the special case of Bayesian pure exchange economy. Due to its specific structure, there are some positive results; see section 10.1.

8.2   *Ex Ante* Solutions

Yannelis (1991) established a private information core allocation existence theorem for an infinite Bayesian pure exchange economy. For the finite economy, an existence theorem can be proved by direct application of Scarf's theorem for nonemptiness of the core (theorem 8.A.2 in the appendix to this chapter) to the non-side-payment game defined by

$$V'(S) := \{u \in \mathbf{R}^N \mid \exists \, x^S \in F'^S : \forall \, j \in S : u_j \leq Eu^j(x^j)\}.$$

Lefebvre (2001) extended Yannelis' (1991) private information core allocation existence theorem to an infinite Bayesian pure exchange economy with *ex ante* non-ordered preference relations.

Ichiishi and Idzik (1996) established an *ex ante* Bayesian incentive-compatible strong equilibrium existence theorem for the Bayesian society: They first established a general existence theorem for the Bayesian societies with externalities in which each utility function u^j depends fully on $(c, t) \in C \times T$. The assumptions in these theorems are stated, however, partly in terms of derivative concepts such as the parameterized non-side-payment games. Then, they derived from the general theorem an existence theorem for the specific class of Bayesian societies without externalities in which each utility function u^j depends only on $(c^j, t) \in C^j \times T$. Assumptions of the theorem for this specific class are stated only in terms of the exogenously given data $\mathcal{S}$. Notice that in spite of the terminology "no-externalities", the feasible-strategy correspondences F^S depend fully on $x \in X$, and to this extent externalities are still considered.

We present the main result of Ichiishi and Idzik (1996) for the no-externality case. Of course, given the no-externality condition, the normal-form games or the Bayesian games are no longer included in the analysis, but economic models extended to cover asymmetric information are included, e.g., the Bayesian pure exchange economy (example 2.2.1), the Bayesian coalition production economy (example 2.2.2), and the Bayesian production economy with interdependent organizations that coexist as firms. Due to the full dependence of F^S on $x \in X$, the core of a production economy with public goods can also be analyzed (given the outsiders' production of public goods specified in $x^{N \setminus S}$ the insiders' feasible strategy set $F^S(x)$ describes the sum of these public goods and the insiders' production possibility set).

A subfamily $\mathcal{B}$ of $\mathcal{N}$ is called *balanced* if there exists $\{\lambda_S\}_{S \in \mathcal{B}} \subset \mathbf{R}_+$ such that $\sum_{S \in \mathcal{B}: S \ni j} \lambda_S = 1$ for every $j \in N$.

THEOREM 8.2.1 (Ichiishi and Idzik, 1996) *Let $\mathcal{S}$ be a Bayesian society in the private information case. Assume that each von Neumann-Morgenstern utility function u^j depends only on $C^j \times T$. Assume also*
(i) *for any j, C^j is a nonempty, compact, convex, and metrizable subset of a Hausdorff locally convex topological vector space over $\mathbf{R}$;*
(ii) *for any j and any t, $u^j(\cdot, t)$ is continuous and linear affine in C^j;*
(iii) *for any S and any t, $\mathbf{C}_0^S(t)$ is nonempty, closed and convex;*
(iv) *for any S, correspondence F^S is both upper and lower semicontinuous in X, and has nonempty, closed and convex values;*
(v) *for any $\bar{x} \in X$ and any balanced family $\mathcal{B}$ with the associated balancing*

8: Existence

coefficients $\{\lambda_S\}_{S\in\mathcal{B}}$, *it follows that*

$$\sum_{S\in\mathcal{B}} \lambda_S \tilde{F}^S(\bar{x}) \subset F^N(\bar{x}),$$

where $\tilde{F}^S(\bar{x}) := \{x \mid x^S \in F^S(\bar{x}), \ x^{N\setminus S} = \mathbf{0}\}$;
(vi) *for any* S, *either* F^S *is a constant correspondence and its value contains a Bayesian incentive-compatible strategy bundle, or for any* S *and any* $\bar{x} \in X$, *there exists* $\hat{x}^S \in F^S(\bar{x})$ *which is strictly Bayesian incentive-compatible, that is, for all* $j \in S$ *and all* $\bar{t}^j, \tilde{t}^j \in T^j$ *for which* $\bar{t}^j \neq \tilde{t}^j$,

$$Eu^j(\hat{x}^j \mid \bar{t}^j) > Eu^j(\hat{x}^j(\tilde{t}^j) \mid \bar{t}^j).$$

Then, there exists an ex ante Bayesian incentive-compatible strong equilibrium of $\mathcal{S}$.

The affine linearity condition (ii) on $u^j(\cdot, t)$ requires some comments: If choices here are interpreted as *pure choices*, then this assumption imposes the strong condition of risk-neutrality on the players' preference relations. If, on the other hand, choices are interpreted as mixed choices, then the utility here should be interpreted as the expected utility. Of course, the expected utility is linear in probabilities, so the assumption is automatically satisfied under the second interpretation of the choices (that is, the affine linearity does not have to be stated as an assumption); see corollary 10.2.1 below.

Condition (v) shows the extent to which the grand coalition is efficient. It also implies that the domain of strategies for each coalition S is the same as the domain of strategies for the grand coalition N, i.e., $T(S) = T(N)$.

Proof of Theorem 8.2.1 We show that the static society

$$(\{X^j, Eu^j\}_{j\in N}, \{\hat{F}^S\}_{S\in\mathcal{N}}, \{\{N\}\})$$

satisfies conditions (i), (ii), (iii), (vi) and (vii) of corollary 8.A.6 of the appendix to this chapter; to avoid confusion here, we call these conditions (A.i), (A.ii), (A.iii), (A.vi) and (A.vii), respectively.

Since each domain $T(S)$ is finite, conditions (A.i), (A.ii) and (A.vii) are immediate consequences of assumptions (i) and (ii) of the present theorem 8.2.1.

Condition (A.vi) is also immediate, from the fact that if $\{x^{(S),j}\}_{j\in S} \in \hat{F}^S(\bar{x})$, in particular, if it is private measurable and Bayesian incentive-compatible, then the strategy bundle $x \in F^N(\bar{x})$ defined by

$$\forall j \in N : x^j := \sum_{S\in\mathcal{B}: S\ni j} \lambda_S x^{(S),j}$$

is also private measurable and Bayesian incentive-compatible (here we use the affine linearity assumption on $u^j(\cdot, t)$).

We show condition (A.iii). The present assumption (vi) guarantees that $\hat{F}^S$ is nonempty-valued. Clearly, it is closed-valued and upper semicontinuous.

To prove lower semicontinuity of $\hat{F}^S$, choose any sequence $\{\bar{x}^{(\nu)}\}_{\nu=1}^{\infty}$ in X which converges to $\bar{x}$, and any $x^S \in \hat{F}^S(\bar{x})$. Choose also $\hat{x}^S \in F^S(\bar{x})$ given in (vi). By lower semicontinuity of F^S, there exist sequences $\{x^{(\nu)S}\}_\nu$ and $\{\hat{x}^{(\nu)S}\}_\nu$ converging to x^S and $\hat{x}^S$, respectively, such that $x^{(\nu)S} \in F^S(\bar{x}^{(\nu)})$ and $\hat{x}^{(\nu)S} \in F^S(\bar{x}^{(\nu)})$ for every ν.

For each $\alpha \in [0, 1]$, define

$$x^S[\alpha] := \alpha x^S + (1-\alpha)\hat{x}^S, \text{ and}$$
$$x^{(\nu)S}[\alpha] := \alpha x^{(\nu)S} + (1-\alpha)\hat{x}^{(\nu)S}.$$

Clearly, $x^{(\nu)S}[\alpha] \in F^S(\bar{x}^{(\nu)})$ for all α. Define

$$\alpha_\nu := \sup\{\alpha \in [0,1] \mid x^{(\nu)S}[\alpha] \in \hat{F}^S(\bar{x}^{(\nu)})\}.$$

It suffices to show that $\alpha_\nu \to 1$ as $\nu \to \infty$.

Due to assumptions (ii) and (vi), and also due to the fact that $x^S \in \hat{F}^S(\bar{x})$, it follows that for all $j \in S$ and all $\bar{t}^j, \tilde{t}^j \in T^j$ for which $\bar{t}^j \neq \tilde{t}^j$,

$$Eu^j(x^S[\alpha], \bar{x}^{N\setminus S} \mid \bar{t}^j) > Eu^j(x^j[\alpha](\tilde{t}^j), x^{S\setminus\{j\}}[\alpha], \bar{x}^{N\setminus S} \mid \bar{t}^j),$$

for all $\alpha \in (0, 1)$.

Now, fix α sufficiently close to 1. For all $j \in S$ and all $\bar{t}^j, \tilde{t}^j \in T^j$ for which $\bar{t}^j \neq \tilde{t}^j$, by the continuity of $u^j(\cdot, t)$,

$$Eu^j(x^{(\nu)S}[\alpha], \bar{x}^{(\nu)N\setminus S} \mid \bar{t}^j) > Eu^j(x^{(\nu)j}[\alpha](\tilde{t}^j), x^{(\nu)S\setminus\{j\}}[\alpha], \bar{x}^{(\nu)N\setminus S} \mid \bar{t}^j),$$

for all $\nu \geq \nu(\alpha)$ and some natural number $\nu(\alpha)$. This implies that $\alpha_\nu \geq \alpha$ for all $\nu \geq \nu(\alpha)$, which establishes lower semicontinuity of $\hat{F}^S$. □

REMARK 8.2.2 The above proof also establishes that the same conditions guarantee the existence of a Bayesian incentive-compatible strong equilibrium of S for the mediator-based approach (so that each strategy in coalition S is T^S-measurable). □

Ever since Ichiishi and Idzik established an earlier version of theorem 8.2.1 in early summer 1991, the need for the affine linearity assumption on utility functions (ii) when Bayesian incentive compatibility is involved had been known in profession, but in a very specific model (like the pure exchange economy) this assumption can be avoided due to the specific structure of the model and the definition of a strategy (see, e.g., theorem 10.2.2).

For an example of a Bayesian society with linear utility functions which does not satisfy the balancedness assumption on feasible-strategy correspondences (v), so has an empty core, see remark 10.2.3.

We present our view on how to evaluate the existence and the nonexistence results. Each economic or game-theoretical model mimics the real world we live in, and each interactive mode specifies players' relationships according to which a game is played. The associated descriptive solution is the outcome that we (analysts) expect to prevail in equilibrium as a result of play of the game, so describes the phenomenon observed in the real world.

If the existence of an equilibrium (e.g., existence of a core allocation or of a Bayesian incentive-compatible strong equilibrium) is guaranteed, we capture the nature of the real phenomena as properties of the equilibrium.

In the case the existence is unlikely, the theorists have held three alternative views on the solution in the past. The first view asserts that the solution is a wrong concept to apply to the real world, so proposes to adopt an alternative solution or even to formulate an alternative model. In the prisoner's dilemma game, for example, a strong equilibrium does not exist. Theorists have sometimes said, "So, the strong equilibrium concept is problematic." Since the unique Nash equilibrium in the same game is not Pareto optimal, theorists have also studied the repeated game (another game) and invoked the folk theorem in order to achieve a Pareto optimal outcome of the original one-shot game.

We contend, however, that the first view does not solve the original problem of understanding the real world; we endorse the following second and the third views. The second view takes the nonexistence result as a warning signal that the model misses important aspects of the reality. To remedy the problem, therefore, we improve the model so that the modified model better reflects the real world and interactive mode.

The third view concerns the situation in which the model captures the essence of the study object, so cannot be improved. Then, we want to analyze a game played within the framework of this model, and not a game in an imaginary world. We also want to study the specific interactive mode that also mimics the pattern of play in the real world. If the associated solution does not exist, we do not apply another solution, since the latter reflects an unrealistic pattern of play. Instead, we accept the fact that the observed phenomena are disequilibrium phenomena, that is, phenomena which we experience in the course of successive formations of blocking coalitions. Thus, if the existence of an equilibrium is not guaranteed, the study object should be the endless formations of coalitions.[1]

---

[1] We quoted the prisoner's dilemma game for the expository purpose. We do not say that this one-shot game mimics the major aspects of the present-day world, or that

8.A   Appendix to Chapter 8

We will review two static game-theoretic models and associated descriptive solution concepts in this appendix. The first model, due to Aumann and Peleg (1960), is called a game in characteristic function form without side payments, or simply a non-side-payment game, and defines the set of attainable utility allocations in each coalition.

Let N be a finite set of players, and let $\mathcal{N} := 2^N \setminus \{\emptyset\}$ be the family of all nonempty coalitions. A non-side-payment game associates with each coalition S a subset $\tilde{V}(S)$ of $\mathbf{R}^S$. The intended interpretation is that the members of S can make a coordinated choice of strategies and realize a utility allocation in S, $\{u_j\}_{j \in S} \in \mathbf{R}^S$, iff $\{u_j\}_{j \in S} \in \tilde{V}(S)$. Notice that the feasible-strategy concept is hidden behind the model, although it is essential in understanding the meaning of the set $\tilde{V}(S)$. Notice also that the attainability of S's utility allocations is determined here independent of strategy choices of the outsiders $N \setminus S$, and this no-externalities simplifies the analysis and at the same time limits the applicability of the model.

By identifying the #S-dimensional space $\mathbf{R}^S$ with the subspace of the #N-dimensional space,

$$\{u \in \mathbf{R}^N \mid \forall\, j \in N \setminus S : u_j = 0\},$$

we can define a non-side-payment game as a correspondence $\tilde{V} : \mathcal{N} \to \mathbf{R}^N$ such that $\tilde{V}(S) \subset \mathbf{R}^S$ for every $S \in \mathcal{N}$. The past studies of a non-side-payment game have proven that we can facilitate our analysis if we look at the cylinder in $\mathbf{R}^N$ based on $\tilde{V}(S)$,

$$V(S) := \{u \in \mathbf{R}^N \mid (\{u_j\}_{j \in S}, \mathbf{0}) \in \tilde{V}(S)\},$$

instead of $\tilde{V}(S)$ itself. We thus have the following definition:

DEFINITION 2.1.2   (Aumann and Peleg, 1960) Let N be a finite player set and let $\mathcal{N} := 2^N \setminus \{\emptyset\}$ be the family of all nonempty coalitions. A *game in characteristic function form without side payments* is a correspondence $V : \mathcal{N} \to \mathbf{R}^N$, such that

$$\forall\, S \in \mathcal{N} : (\forall\, u, v \in \mathbf{R}^N : u_j = v_j \text{ for all } j \in S) :$$
$$[u \in V(S) \Leftrightarrow v \in V(S)].$$

It is synonymously called a *non-side-payment game*, a *non-transferable-utility game*, or an *NTU game*.

---

the associated repeated game is unrealistic. On the contrary, the implications of long-run threat and commitment in the repeated game, as formalized in the folk theorem, constitute a real principle working in the present-day world.

8: Existence

The central descriptive solution of game V is a core utility allocation, defined as a utility allocation which is attainable in the grand coalition N (condition (i) in the following definition 8.A.1), and which cannot be improved upon by any coalition (condition (ii)).

DEFINITION 8.A.1 (Aumann and Peleg, 1960) Let $V : N \to \mathbf{R}^N$ be a non-side-payment game. The *core* of the game V is the set of utility allocations $u^* \in \mathbf{R}^N$ such that
(i) $u^* \in V(N)$, and
(ii) it is not true that

$$\exists S \in \mathcal{N} : \exists u \in V(S) : \forall j \in S : u_j > u_j^*.$$

To date, there are two fundamental theorems, each establishing a condition for nonemptiness of the core of a non-side-payment game: balancedness and ordinal convexity. In this book, we concentrate on the former. For each $S \subset N$, define the characteristic vector of S, $\chi_S \in \mathbf{R}^N$, by:

$$(\chi_S)_j := \begin{cases} 1 & \text{if } j \in S \\ 0 & \text{if } j \in N \setminus S. \end{cases}$$

A subfamily $\mathcal{B}$ of $\mathcal{N}$ is called *balanced*, if there exists an indexed set of nonnegative real numbers, $\{\lambda_S \in \mathbf{R}_+ \mid S \in \mathcal{B}\}$, such that $\sum_{S \in \mathcal{B}} \lambda_S \chi_S = \chi_N$. Notice that

$$\sum_{S \in \mathcal{B}} \lambda_S \chi_S = \chi_N \Leftrightarrow \forall j \in N : \sum_{S \in \mathcal{B}: S \ni j} \lambda_S = 1.$$

The set $\{\lambda_S \in \mathbf{R}_+ \mid S \in \mathcal{B}\}$ is called *associated balancing coefficients*.

A partition is a balanced family for which the associated balancing coefficients are all 1. A non-trivial, balanced family for the three-person case ($N = \{1,2,3\}$) is $\mathcal{B} = \{\{1,2\}, \{2,3\}, \{3,1\}\}$, for which the associated balancing coefficients are given by $\lambda_S = 1/2$ for all $S \in \mathcal{B}$.

Here is one interpretation of the balancedness condition on a family $\mathcal{B}$: Suppose that each player can put fractions of his total effort into several coalitions. Nobody puts positive effort in coalitions other than those of $\mathcal{B}$. Player j puts fraction λ_S of his total effort into coalition $S \in \mathcal{B}$ for which $S \ni j$. Since the fractions are summed up to 1, $\sum_{S \in \mathcal{B}: S \ni j} \lambda_S = 1$ for every player j. Every member of coalition S puts fraction λ_S of his total effort.

A non-side-payment game V is called *balanced*, if for every balanced family $\mathcal{B}$, it follows that

$$\bigcap_{S \in \mathcal{B}} V(S) \subset V(N).$$

The balancedness condition on non-side-payment game V says that for any balanced family $\mathcal{B}$, if utility allocation $u \in \mathbf{R}^N$ is such that $\{u_j\}_{j \in S}$ is attainable in each member S of $\mathcal{B}$, then this allocation u is attainable in the grand coalition N. This way, the balancedness condition on V makes explicit the extent to which the grand coalition is efficient.

The balancedness condition on game V is best interpreted in each economic context: In an specific economic context, one starts with an economic model $\mathcal{E}$ which idealizes the context, and derive from $\mathcal{E}$ the associated non-side-payment game V. Cooperative behavior in $\mathcal{E}$ is summarized as a cooperative behavior in V. The art of theory is to identify economically meaningful conditions on $\mathcal{E}$ under which the associated game V is proved to be balanced.

Scarf established the following theorem 8.A.2, which says that the balancedness condition on game V, together with minor regularity conditions, guarantees nonemptiness of the core of V. Condition (i) of the theorem, customarily called the comprehensiveness, says that if utility allocation u^S is feasible in coalition S and if $v^S \leq u^S$, then utility allocation v^S is also feasible in coalition S; this is nothing but free disposal. Condition (ii) says that there is an upper bound for the feasible, individually rational utility allocations. Condition (iii) is a technical assumption; actually, it can be replaced by a weaker assumption: $V(N)$ is closed in $\mathbf{R}^N$.

THEOREM 8.A.2 (Scarf, 1967) *Let $V : \mathcal{N} \to \mathbf{R}^N$ be a non-side-payment game, and define $b \in \mathbf{R}^N$ by: $b_j := \sup \{u_j \in \mathbf{R} \mid u \in V(\{j\})\}$, for every $j \in N$. The core of V is nonempty if:*
(i) $V(S) - \mathbf{R}_+^N = V(S)$ *for every* $S \in \mathcal{N}$;
(ii) *There exists $M \in \mathbf{R}$ such that for every $S \in \mathcal{N}$, $[u \in V(S), u \geq b]$ implies $[u_j < M$ for every $j \in S]$;*
(iii) $V(S)$ *is closed in $\mathbf{R}^N$ for every $S \in \mathcal{N}$; and*
(iv) V *is balanced.*

It is easy to extend the model of non-side-payment game, the core concept, and the Scarf theorem, in order to explain endogenous realization of a coalition structure in equilibrium (rather than realization of the grand coalition). However, some coalition structures may not be realized in equilibrium, due to, e.g., a legal restriction. Therefore, let $\mathcal{T}_0$ be the family of admissible coalition structures, *a priori* given to the model. The *generalized core* of game $(V, \mathcal{T}_0)$ is a pair $(u^*, \mathcal{T}^*)$ of a utility allocation $u^* \in \mathbf{R}^N$ and an admissible coalition structure $\mathcal{T}^* \in \mathcal{T}_0$, which satisfies the definition 8.A.1 of the core, except that condition (i) be replaced by:

$$\forall T \in \mathcal{T}^* : u^* \in V(T).$$

The coalitions in $\mathcal{T}^*$ are formed and coexist in equilibrium.

A generalized core is nonempty, if game $(V, \mathcal{T}_0)$ satisfies all the conditions of theorem 8.A.2, except that condition (iv) be replaced by:

$$(\forall \mathcal{B} : \text{balanced}) : \bigcap_{S \in \mathcal{B}} V(S) \subset \bigcup_{T \in \mathcal{T}_0} \bigcap_{T \in \mathcal{T}} V(T).$$

We turn to the second model. It explicitly formulates the strategy concept in the model, and encompasses situations in which the set of attainable utility allocations of coalition S is influenced by strategy-choice of the outsiders $N \setminus S$.

Among the game-theoretical models designed for explicit analysis of players' strategic interaction are: a game in normal form, a game in extensive form, and a Bayesian game (definition 2.1.1). A Bayesian game is an extension of a normal-form game to incomplete information, formulated as an extensive game with a specific context. A *game in normal form* is a specified list of data $\{X^j, u^j\}_{j \in N}$ of a finite player set N, a strategy set X^j, and a utility function $u^j : \prod_{i \in N} X^i \to \mathbf{R}$ for each player j. A *Nash equilibrium* is a descriptive solution to a normal-form game, defined as a strategy bundle $\{x^{*j}\}_{j \in N} \in \prod_{j \in N} X^j$ such that

$$\neg \exists j \in N : \exists x^j \in X^j : u^j(x^j, x^{*N\setminus\{j\}}) > u^j(x^*).$$

It is a noncooperative solution concept: No player is aggressive enough to make a coordinated choice of strategies with other players in order to pursue his self-interest.

The theme of this book is analysis of players' behavior which is aggressive enough to take advantage of the effects of coordinated behavior with other players. For each coalition $S \in \mathcal{N}$, the members' joint strategy space is the Cartesian product, $X^S := \prod_{j \in S} X^j$. Set $X := X^N$ for convenience.

We introduce three ingredients to the normal-form game. The first ingredient is the feasibility concept for strategy bundles. The feasible-strategy correspondence of coalition S is a correspondence

$$F^S : X \to X^S$$

(we may assume that F^S depends only upon $\bar{x}^{N\setminus S} \in X^{N\setminus S}$). By choosing strategies $\bar{x}^{N\setminus S}$, the players outside coalition S influence the members of S, indirectly by restricting S's feasible joint-strategies to $F^S(\bar{x})$. Notice the possibility, $F^S(\bar{x}) \neq \prod_{j \in S} F^{\{j\}}(\bar{x})$; an important possibility not covered by the normal-form game.

The second ingredient generalizes the utility function u^j, in order to cover the situation in which player j's utility also depends on the coalition that j belongs to, as it is affected by a (nonstrategic) environment specific to S. We introduce utility functions,

$$u^j_S : X \to \mathbf{R}, \text{ for every } S \text{ for which } S \ni j.$$

The value $u^j_S(x)$ is player j's utility level when he belongs to coalition S, the outsiders choose strategy bundle $x^{N \setminus S}$, and the members of S choose x^S. As in the normal-from game, the players outside coalition S influence the members of S, directly by affecting the utility function u^j_S of every $j \in S$.

The third ingredient is admissibility of a realized coalition structure. Let $\mathcal{T}_0$ be the family of admissible coalition structures, *a priori* given to the model.

DEFINITION 8.A.3 (Ichiishi, 1981) A *society* is a list of specified data,

$$\mathcal{S} := (\{X^j\}_{j \in N}, \{F^S\}_{S \in \mathcal{N}}, \{u^j_S\}_{j \in S \in \mathcal{N}}, \mathcal{T}_0),$$

where N is a finite player set, X^j is player j's strategy set, feasible or infeasible, $F^S : X \to X^S$ is coalition S's feasible-strategy correspondence, $u^j_S : X \to \mathbf{R}$ is j's utility function as a member of S, and $\mathcal{T}_0$ is a family of admissible coalition structures.

The *strong equilibrium* of a normal-form game is a strategy bundle $x^* \in X$ such that

$$\neg \; \exists \; S \in \mathcal{N} : \exists \; x^S \in X^S : \forall \; j \in S : u^j(x^S, x^{*N \setminus S}) > u^j(x^*).$$

This is readily extended to a society. In the following definition 8.A.4, condition (i) stipulates feasibility of strategy bundle x^* via coalition structure $\mathcal{T}^*$, and condition (ii) stipulates coalitional stability.

DEFINITION 8.A.4 (Ichiishi, 1981) A *social coalitional equilibrium* of society $\mathcal{S}$ is a pair of a strategy bundle and an admissible coalition structure $(x^*, \mathcal{T}^*) \in X \times \mathcal{T}_0$ such that
(i) $\forall \; T \in \mathcal{T}^* : x^{*T} \in F^T(x^*)$; and
(ii) it is not true that

$$\exists \; S \in \mathcal{N} : \exists \; x^S \in F^S(x^*) :$$
$$\forall \; j \in S : \; u^j_S(x^S, x^{*N \setminus S}) > u^j_{T(j)}(x^*),$$

where $T(j)$ is the unique member in $\mathcal{T}^*$ such that $T(j) \ni j$.

8: Existence

Notice that society $\mathcal{S}$ reduces to a game in normal form, and a social coalitional equilibrium reduces to a Nash equilibrium, if the finest coalition structure is the only admissible coalition structure,

$$\mathcal{T}_0 = \{\{\{j\} \mid j \in N\}\},$$

the feasible-strategy correspondences do not impose restrictions,

$$\forall\, S \in \mathcal{N} : \forall\, \bar{x} \in X : F^S(\bar{x}) = X^S,$$

and players receive a very low utility as members of a non-singleton coalition,

$$(\forall\, S \in \mathcal{N} : \#S \geq 2) : \forall\, j \in S : \forall\, x \in X : u_S^j(x) = -\infty,$$

so that there is no point in forming a non-singleton coalition.

Notice also that society $\mathcal{S}$ reduces to a non-side-payment game, and the set of social coalitional equilibria reduces to the core, if the coarsest coalition structure is the only admissible coalition structure,

$$\mathcal{T}_0 = \{\{N\}\},$$

each player's strategy is his utility level (a real number), the feasible-strategy correspondences are the constant correspondences that describe sets of attainable utility allocations,

$$\forall\, S \in \mathcal{N} : \forall\, \bar{x} \in X : F^S(\bar{x}) = \tilde{V}(S),$$

and each player's utility function is the projection of the utility bundle to his own utility,

$$\forall\, S \in \mathcal{N} : \forall\, j \in S : \forall\, u \in \mathbf{R}^N : u_S^j(u) = u_j.$$

We have pointed out that a social coalitional equilibrium of a society is an extension of a strong equilibrium of a game in normal form. Indeed, society $\mathcal{S}$ reduces to a game in normal form, and a social coalitional equilibrium reduces to a strong equilibrium, if the coarsest coalition structure is the only admissible coalition structure,

$$\mathcal{T}_0 = \{\{N\}\},$$

the feasible-strategy correspondences do not impose restrictions,

$$\forall\, S \in \mathcal{N} : \forall\, \bar{x} \in X : F^S(\bar{x}) = X^S,$$

and players' utility functions are not affected by (nonstrategic) environments specific to the coalitions they belong to,

$$\forall\, j \in N : \exists\, u^j : X \to \mathbf{R} : (\forall\, S \in \mathcal{N} : S \ni j) : u_S^j = u^j.$$

It has been widely discussed that a strong equilibrium of a normal-form game frequently fails to exist, as the prisoner's dilemma shows. In the rest of this appendix, however, we will present some positive existence results for a social coalitional equilibrium. The general existence theorem is applicable to a normal-form game, but in such a case the conditions for the existence are too stringent (certainly the prisoner's dilemma game does not satisfy them). The conditions for the existence of a social coalitional equilibrium turn out to be plausible when the model of society is specialized differently. Many interesting economic examples which involve coalition formations fall into the latter category.

The general existence theorem (theorem 8.A.5) imposes conditions not only on the primitive data of society $\mathcal{S}$ but also on the following derivative concept of parameterized non-side-payment games. Here, a strategy bundle $\bar{x} \in X$ is taken as a parameter.

$$V_{\bar{x}}(S) := \left\{ u \in \mathbf{R}^N \;\middle|\; \begin{array}{l} \exists\, x^S \in F^S(\bar{x}) : \forall\, j \in S : \\ u_j \leq u_S^j(x^S, \bar{x}^{N \setminus S}) \end{array} \right\}.$$

A social coalitional equilibrium strategy bundle x^* is a fixed point of the core-correspondence,

$$\bar{x} \mapsto \left\{ \begin{array}{l} \text{the strategy bundles that give rise to the generalized} \\ \text{core utility allocations of game } (V_{\bar{x}}, \mathcal{T}_0) \end{array} \right\}.$$

The Scarf theorem (theorem 8.A.2) is not applicable here, since this correspondence is in general disconnected-set-valued, so even if we can apply the Scarf theorem to each parameterized game $(V_{\bar{x}}, \mathcal{T}_0)$, we cannot obtain a fixed point.

THEOREM 8.A.5 (Ichiishi, 1981) *Let $\mathcal{S}$ be a society. There exists a social coalitional equilibrium of $\mathcal{S}$, if*
(i) *X^j is a nonempty, compact, convex subset of a Hausdorff locally convex topological vector space over $\mathbf{R}$ for every $j \in N$;*
(ii) *u_S^j is continuous in X for every $S \in \mathcal{N}$ and $j \in S$;*
(iii) *F^S is upper and lower semicontinuous in X, and is nonempty- and closed-valued for every $S \in \mathcal{N}$;*
(iv) *given any $\bar{x} \in X$, the non-side-payment game with admissible coalition structures $(V_{\bar{x}}, \mathcal{T}_0)$ is balanced, that is, for every balanced family $\mathcal{B}$,*

$$\bigcap_{S \in \mathcal{B}} V_{\bar{x}}(S) \subset \bigcup_{T \in \mathcal{T}_0} \bigcap_{T \in \mathcal{T}} V_{\bar{x}}(T); \quad \text{and}$$

(v) *given any $\bar{x} \in X$ and any generalized core utility allocation $\bar{u}$ of game $(V_{\bar{x}}, \mathcal{T}_0)$, the set*

$$\bigcup_{T \in \mathcal{T}_0} \prod_{T \in \mathcal{T}} \{x^T \in F^T(\bar{x}) \mid \forall\, j \in T : u^j_T(x^T, \bar{x}^{N \setminus T}) \geq \bar{u}_j\}$$

is convex.

The balancedness assumption (iv) and the convexity assumption (v) in theorem 8.A.5 are probably too obscure to capture, since each imposes on the model an implicit relation between F^S and u^j_S. The following corollary is proved by showing that these assumptions are derived from a plausible assumption only on F^S and another plausible assumption only on u^j_S. This last assumption on u^j_S defines an important special case of the society; the special case is distinct from the game in normal form. Define

$$\tilde{F}^S(\bar{x}) := \{(x^S, \mathbf{0}) \mid x^S \in F^S(\bar{x})\},$$

where $\mathbf{0}$ is the origin of the vector space spanned by $X^{N \setminus S}$, so that $\tilde{F}^S(\bar{x})$ lies in the vector space spanned by X for all S.

COROLLARY 8.A.6 (Ichiishi, 1981) *Let $\mathcal{S}$ be a society. There exists a social coalitional equilibrium of $\mathcal{S}$, if assumptions (i), (ii) and (iii) of theorem 8.A.5 are satisfied, and if*
(vi) *for any $\bar{x} \in X$, and any balanced family $\mathcal{B}$ with associated balancing coefficients $\{\lambda_S\}_{S \in \mathcal{B}}$, it follows that*

$$\sum_{S \in \mathcal{B}} \lambda_S \tilde{F}^S(\bar{x}) \subset \bigcup_{T \in \mathcal{T}_0} \sum_{T \in \mathcal{T}} \tilde{F}^T(\bar{x}); \text{ and}$$

(vii) *for each $j \in N$, there is a quasi-concave utility function $u^j : X^j \to \mathbf{R}$, such that*

$$u^j_S(x) = u^j(x^j), \text{ for all } S \in \mathcal{N}, \; j \in S, \text{ and } x \in X.$$

The special case of a society, as stipulated by condition (vii) of corollary 8.A.6 cannot include the normal-form games. In particular, unlike the normal-from game, the players outside coalition S cannot influence the members of S directly through the utility functions; this is a society in no-externality case. Nevertheless, by choosing strategies $\bar{x}^{N \setminus S}$, the outsiders $N \setminus S$ influence the insiders S, indirectly by restricting S's feasible joint-strategies to $F^S(\bar{x})$. There are abundance of economic examples of the latter kind of externalities, e.g., marketing strategies of competing firms in a production economy in which a coalition of human-resource holders is identified with a firm, public goods, and water pollution to the fishing industry.

We close the appendix to chapter 8 by presenting a variant of the model of society. It is a family of societies, $\mathcal{S}_q$, parameterized by $q \in Q$. For each parameter q, the player set of the society $\mathcal{S}_q$ is N, player j's strategy space is X^j, and his utility function is $u^j(q,\cdot) : X^j \to \mathbf{R}$, the feasible-strategy correspondence of coalition S is $F^S(q,\cdot) : X \to X^S$, and the family of admissible coalition structures is $\mathcal{T}_0$. The new ingredient is that the parameter responds to a prevailing value of the pair, $(\bar{q}, \bar{x})$, and this response is given by correspondence,

$$G : Q \times X \to Q.$$

The intended scenario goes as follows: We start with a parameter $\bar{q}$, so the players interact with each other in the society $\mathcal{S}_{\bar{q}}$. If they end up choosing a strategy bundle $\bar{x}$ in $\mathcal{S}_{\bar{q}}$, the parameter changes from $\bar{q}$ to a new value $\bar{\bar{q}} \in G(\bar{q}, \bar{x})$, and the players re-start their interaction in the new society $\mathcal{S}_{\bar{\bar{q}}}$.

DEFINITION 8.A.7 A *parameterized family of societies* is a list of specified data,

$$(\{\mathcal{S}_q\}_{q \in Q},\ G) := (\{X^j, u^j\}_{j \in N},\ \{F^S\}_{S \in \mathcal{N}},\ Q,\ G,\ \mathcal{T}_0),$$

where N is a finite player set, Q is a parameter set, X^j is player j's strategy set, feasible or infeasible, $u^j : Q \times X^j \to \mathbf{R}$ is j's utility function, $F^S : Q \times X \to X^S$ is coalition S's feasible-strategy correspondence, $G : Q \times X \to Q$ is parameter's response correspondence, and $\mathcal{T}_0$ is a family of admissible coalition structures.

The social coalitional equilibrium concept is readily extended to the parameterized family of societies. It is a triple of a parameter value, a strategy bundle and an admissible coalition structure, $(q^*, x^*, \mathcal{T}^*) \in Q \times X \times \mathcal{T}_0$, such that (1) given q^*, $(x^*, \mathcal{T}^*)$ is a social coalitional equilibrium of the society $\mathcal{S}_{q^*}$, and (2) given (q^*, x^*), q^* is stationary.

DEFINITION 8.A.8 A *social coalitional equilibrium* of a parameterized family of societies $(\{\mathcal{S}_q\}_{q \in Q},\ G)$ is a triple $(q^*, x^*, \mathcal{T}^*) \in Q \times X \times \mathcal{T}_0$ such that
(i) $\forall\ T \in \mathcal{T}^* : x^{*T} \in F^T(q^*, x^*)$;
(ii) $\neg\ \exists\ S \in \mathcal{N} : \exists\ x^S \in F^S(q^*, x^*) : \forall\ j \in S : u^j(q^*, x^j) > u^j(q^*, x^{*j})$;
(iii) $q^* \in G(q^*, x^*)$.

THEOREM 8.A.9 Let $(\{\mathcal{S}_q\}_{q \in Q},\ G)$ be a parameterized family of societies. There exists a social coalitional equilibrium of $(\{\mathcal{S}_q\}_{q \in Q},\ G)$, if
(i) X^j is a nonempty, convex, compact subset of $\mathbf{R}^{m_j}$ for every $j \in N$;
(i') Q is a nonempty, convex, compact subset of $\mathbf{R}^{m_0}$;
(ii) u^j is continuous in $Q \times X^j$, for every $j \in N$;

(iii) F^S *is upper and lower semicontinuous in* $Q \times X$, *and is nonempty- and closed-valued, for every* $S \in \mathcal{N}$;

(iii′) G *is upper semicontinuous in* $Q \times X$, *and is nonempty- closed- and convex-valued;*

(iv) *for any* $(\bar{q}, \bar{x}) \in Q \times X$, *the non-side-payment game* $V_{\bar{q},\bar{x}} : \mathcal{N} \to \mathbf{R}^N$ *defined by*

$$V_{\bar{q},\bar{x}}(S) := \{u \in \mathbf{R}^N \mid \exists\, x^S \in F^S(\bar{q},\bar{x}) : \forall\, j \in S : u_j \leq u^j(\bar{q}, x^j)\}$$

is balanced; and

(v) *for any* $(\bar{q}, \bar{x}) \in Q \times X$ *and any utility allocation* $\bar{u}$ *in the core of* $V_{\bar{q},\bar{x}}$, *the set*

$$\{x \in F^N(\bar{q},\bar{x}) \mid \forall\, j \in N : u^j(\bar{q}, x^j) \geq \bar{u}_j\}$$

is convex.

Chapter 9

Approaches to Information Revelation

Each player j is endowed with his private information structure T^j, so he knows his true type $\bar{t}^j$ at the beginning of the *interim* period. By the time the strategy execution is over, player j will have narrowed down the range of his colleague i's possible true types to a subset A_i^j of T^i. In other words, while the players start with the null communication system $\{T^j\}_{j\in N}$, they end up with an endogenously determined finer communication system $\{A^j\}_{j\in N}$. This information revelation process is not easy to analyze, since a player j may not want to pass on his private information to his colleagues, and even if j decides to do so, his colleagues may think that j is not truthfully passing on his information but is trying to manipulate them with false information. This chapter will review two approaches taken in the literature for endogenous determination of an information structure: (1) passive information revelation by action; and (2) active information revelation by credible transmission of information (e.g., by credible talking). Approach (1) is classified into two specific approaches: (1a) information revelation by contract execution, and (1b) information revelation by choosing a contract. There are formal works in the literature for approaches (1a) and (2); we will review the works for (1a) in sections 9.1 and 9.2, and the works for (2) in section 9.5. Formal study of approach (1b) is still left uncultivated, but we will present known examples to illustrate the idea in section 9.3. One formal work pursuing approach (1b) is an application of the (trembling-hand) perfect equilibrium or the sequential equilibrium; although no general theorems have been obtained to date, we will present this line of thoughts in section 9.4, adapted to our cooperative framework.

9.1 By Contract Execution

This approach borrows ideas for information update from the rational expectations equilibrium analysis. In the latter framework, somebody announces a price function $\mathbf{p} : T \to \mathbf{R}^l$ to the economic agents. When an economic agent observes a price vector p, he realizes that the event $\mathbf{p}^{-1}(p) \subset T$ has occurred (see, e.g., Radner (1979)). In the Bayesian society in the private information case, the members of coalition S agree on a strategy bundle x^S, everybody in the coalition knows his colleague i's strategy x^i, so if i makes a choice c^i then the members of S realize that the event $\left(x^i\right)^{-1}(c^i) \subset T$ has occurred. Choice is postulated to be observable, so moral hazard problems are excluded.

We remark that while formation of the price function $\mathbf{p}$ is still an open research agenda, there is an explicit scenario for formation of the strategy bundle x^S, that is, coordinated strategy-choice by the members of S.

This section and next section present two works on this information revelation process, Ichiishi, Idzik and Zhao (1994), and Ichiishi and Radner (1999). The essential message of these works is that even if a game starts with the situation characterized as the private information case, it ends up with the full communication system.

Ichiishi, Idzik and Zhao (1994) studied an *ex ante* determination of a strategy bundle, taking into account the above process in the general framework of the Bayesian society

$$\mathcal{S} := \left(\{C^j, T^j, u^j, \{\pi^j(\cdot \mid t^j)\}_{t^j \in T^j}\}_{j \in N}, \ \{\mathbf{C}_0^S, T(S), F^S\}_{S \in \mathcal{N}} \right)$$

(definition 2.1.3). Since the focus here is an *ex ante* equilibrium, we postulate that there exists an *ex ante* probability π on T, assumed to be objective (therefore, $\pi^j(\cdot \mid t^j)$ is the conditional probability of π given t^j). Moreover, we assume $\pi \gg \mathbf{0}$ (hence $T(S) = T$) for simplicity of the analysis; this assumption may be dropped.

For full analysis of players' behavior before and after update of information, Ichiishi, Idzik and Zhao introduced an additional structure to $\mathcal{S}$ (postulates 9.1.1 and 9.1.2 below). The first postulate says that each player makes choice twice, once in the first *interim* period, and then in the second *interim* period. The players act simultaneously at each round.

POSTULATE 9.1.1 For each player $j \in N$, his choice set is of the form, $C^j = C_1^j \times C_2^j$.

Set $c^j = (c_1^j, c_2^j)$. Player j makes choice c_1^j (c_2^j, resp.) at his information set of the first *interim* period (the second *interim* period, resp.).

9: Information Revelation

For two information structures $\mathcal{B}$ and $\mathcal{C}$ (algebras on T), denote by $\mathcal{B} \bigvee \mathcal{C}$ the algebra generated by $\mathcal{B} \bigcup \mathcal{C}$ (the smallest algebra on T that contains both $\mathcal{B}$ and $\mathcal{C}$). For any set Z and any function $f : T \to Z$, denote by $\mathcal{A}(f)$ the algebra generated by f (the smallest algebra on T that contains the sets $\{f^{-1}(z) \mid z \in Z\}$).

A strategy of player j in coalition S is also denoted by $(x_1^j(\cdot), x_2^j(\cdot))$. Given a strategy bundle x^S, information is processed within coalition S in the following way: In the first *interim* period, each player has only his own private information. So, the component $x_1^j(\cdot)$ has to be $\mathcal{T}^j$-measurable. *If it is common knowledge in S that player j has the incentive to make a choice (say, c_1^j) in the first interim period according to his true type*, then by the beginning of the second *interim* period the occurrence of event

$$\{t^j \in T^j \mid x_1^j(t^j) = c_1^j\}$$

has become common knowledge in S. Let $\bar{t}$ be the true type profile, and suppose choice bundle $c_1^S \in C_1^S$ is made in the first *interim* period. Then each player j has the information that event

$$\{t \in T \mid t^j = \bar{t}^j,\ x_1^S(t) = c_1^S\}$$

has occurred with probability 1.

When designing the other component of the strategy bundle $x_2^S = (x_2^i)_{i \in S}$, the members can anticipate that the information structure,

$$\hat{\mathcal{T}}^i(x_1^S) := \mathcal{T}^i \bigvee \left(\bigvee_{j \in S} \mathcal{A}(x_1^j) \right),$$

is available to i at the beginning of the second *interim* period, and make each x_2^i measurable with respect to it. Thus, we can make the following postulate of *information-revelation process*:

POSTULATE 9.1.2 Given any strategy bundle $\bar{x} \in X$, coalition S designs only those $x^S \in F^S(\bar{x})$ such that for all $j \in S$ it follows that
(i) x_1^j is measurable with respect to $\mathcal{T}^j$, and
(ii) x_2^j is measurable with respect to $\hat{\mathcal{T}}^j(x_1^S)$.

Denote by $F'^S(\bar{x})$ the set of those feasible strategies x^S that satisfy postulate 9.1.2 (information-revelation process):

$$F'^S(\bar{x}) := \left\{ x^S \in F^S(\bar{x}) \;\middle|\; \begin{array}{l} \forall\, j \in S: \\ x_1^j \text{ is measurable with respect to } \mathcal{T}^j,\text{ and} \\ x_2^j \text{ is measurable with respect to } \hat{\mathcal{T}}^j(x_1^S) \end{array} \right\}.$$

Recall that in order for the present information-revelation process to work, the members of coalition S need to have the common knowledge that each player has the incentive to make a choice in the first *interim* period according to his true type. After all, the contract will not be enforced, if some member has the incentive to make a choice with false pretension about his true type either during the first *interim* period or during the second *interim* period. If the members of S foresee at the time of contract design that a particular contract x^S may later induce such false pretension, they do not agree on the contract x^S. Instead of the strategy set $F'^S(\bar{x})$, therefore, they will consider only the restricted subset $\hat{F}^S(\bar{x})$ of those strategy bundles that are Bayesian incentive-compatible. The definition of Bayesian incentive compatibility in the context of postulates 9.1.1 and 9.1.2 is much involved; see postulate 9.A.1 and fact 9.A.2 in the appendix to this chapter for the detail. With the feasible-strategy correspondences $\hat{F}^S : X \to X^S$ modified this way, we can define a *Bayesian incentive-compatible strong equilibrium* as in definition 5.2.2:
(i) $x^* \in \hat{F}^N(x^*)$; and
(ii) it is not true that

$$\exists\, S \in \mathcal{N} : \exists\, x^S \in \hat{F}^S(x^*) : \forall\, j \in S : Eu^j(x^S, x^{*N\setminus S}) > Eu^j(x^*).$$

The difficulty in establishing an existence theorem for the present Bayesian incentive-compatible strong equilibrium lies in the fact that even if F^S is well-behaved, $\hat{F}^S$ is not convex-valued or upper semicontinuous, so the standard existence techniques do not apply. Ichiishi, Idzik and Zhao (1994) provided generic existence theorems for a Bayesian incentive-compatible strong equilibrium. Again, a generic existence theorem was established first for the general class of Bayesian societies with externalities in which each utility function u^j depends fully on $(c, t) \in C \times T$, and then as its application a generic existence theorem was established for the specific class of Bayesian societies without externalities[1] in which each utility function u^j depends only on $(c^j, t^j) \in C^j \times T^j$. Assumptions of the theorem for this specific class are stated only in terms of the exogenously given data $\mathcal{S}$'s. Notice again that in spite of the terminology "no-externalities", the feasible-strategy correspondences F^S depend fully on $x \in X$, and to this extent externalities are still considered.

A Bayesian society studied here is a specified list of data,

$$\mathcal{S} := (\{C^j, T^j, u^j, \pi\}_{j \in N}, \{\mathbf{C}_0^S, F^S\}_{S \in \mathcal{N}})$$

(definition 2.1.3), with a rich structure (postulates 9.1.1 and 9.1.2). Here, the *ex ante* objective probability π is assumed to be strictly positive.

---
[1] The concept of no-externalities is more stringent here than in theorem 8.2.1.

9: Information Revelation

We first discuss the no-externalities case: $u^j(x,t) = u^j(x^j, t^j)$. One part of the data ($\{C^j, T^j, u^j, \pi\}_{j \in N}$, $\{\mathbf{C}_0^S\}_{S \in \mathcal{N}}$) will be fixed throughout. By changing the other part of the data from $\{F^S\}_{S \in \mathcal{N}}$ to $\{F^{\dagger S}\}_{S \in \mathcal{N}}$, we obtain *another* Bayesian society

$$\mathcal{S}^\dagger := \left(\{C^j, T^j, u^j, \pi\}_{j \in N},\ \{\mathbf{C}_0^S, F^{\dagger S}\}_{S \in \mathcal{N}} \right).$$

By varying $\{F^S\}_{S \in \mathcal{N}}$, one obtains the *space* of Bayesian societies, SPACE_{ne}. The space will be endowed with a natural pseudo-metric d. The pseudo-metric space ($\text{SPACE}_{\text{ne}}, d$) of Bayesian societies will thus be constructed.

A property $\mathcal{P}$ is called a *generic property* of a Bayesian society *in* SPACE_{ne}, if there exists an open and dense subset $\text{SPACE}'_{\text{ne}}$ of ($\text{SPACE}_{\text{ne}}, d$) such that every $\mathcal{S} \in \text{SPACE}'_{\text{ne}}$ satisfies $\mathcal{P}$. The first main theorem of this section (theorem 9.1.4 below) clarifies conditions on SPACE_{ne} under which the following are generic properties of a Bayesian society:

- There exist multitude of Bayesian incentive-compatible strong equilibria.

- There exists a Bayesian incentive-compatible strong equilibrium x^* such that it fully reveals private information by the end of the first interim period.

Thus, a Bayesian society generically has a Bayesian incentive-compatible strong equilibrium which processes the null communication system to the full communication system through players' actions during the first *interim* period.

In the context of the present structure (postulates 9.1.1 and 9.1.2), the essential role of the feasible-strategy correspondences $\{F^S\}_{S \in \mathcal{N}}$ in Bayesian society $\mathcal{S}$ is described by the correspondences $\{G^S\}_{S \in \mathcal{N}}$ defined by

$$G^S(\bar{x}) := \left\{ x^S \in F^S(\bar{x}) \ \middle| \ \begin{array}{l} \forall j \in S: \\ x_1^j \text{ is } \mathcal{T}^j\text{-measurable, and} \\ x_2^j \text{ is } \mathcal{T}^S\text{-measurable} \end{array} \right\}.$$

We will give a precise definition of set SPACE_{ne}: It consists of all Bayesian societies satisfying the five conditions in assumption 9.1.3. The first two conditions are on the fixed data ($\{C^j, T^j, u^j, \pi\}_{j \in N}, \{C_0^S\}_{S \in \mathcal{N}}$), hence on the fixed strategy-spaces $X^j := \{x^j : T \to C^j\}$. Conditions (iii)-(v) are on each $\{F^S\}_S$ which defines a member of SPACE_{ne}.

ASSUMPTION 9.1.3 (i) (1) For every $j \in N$, his choice set C^j is a nonempty, compact, convex and metrizable subset of a real Hausdorff locally convex topological vector space. (2) For every $j \in N$, his von

Neumann-Morgenstern utility function u^j depends only on $(c^j, t^j) \in C^j \times T^j$, and moreover, $u^j(\cdot, t^j)$ is linear affine and continuous on C^j for each t^j.

(ii) For each $j \in N$, there exist $c_m^j \in C^j$ and a finite subset C_f^j of C^j such that (1) for all $c^j \in C_f^j$ and all $t^j \in T^j$, $u^j(c^j, t^j) > u^j(c_m^j, t^j)$; (2) for all $c^j, c'^j \in C_f^j$ for which $c^j \neq c'^j$, it follows that $c_1^j \neq c_1'^j$; (3) for all $t^j, t'^j \in T^j$ for which $t^j \neq t'^j$, there exist $c^j, c'^j \in C_f^j$ such that $u^j(c^j, t^j) > u^j(c'^j, t^j)$, and $u^j(c'^j, t'^j) > u^j(c^j, t'^j)$.

(iii) (1) For each $S \in \mathcal{N}$, correspondence $G^S : X \to C^S$ is upper and lower semicontinuous in X, and for each $\bar{x} \in X$, $G^S(\bar{x})$ is nonempty, closed and convex. (2) The correspondence $G^N(\cdot)$ is a constant correspondence on X, so one may write $G^N := G^N(\bar{x})$. The set G^N is relatively strictly convex (the strict convex combinations of any two distinct members of G^N are in the relative interior of G^N). There exist $x, x' \in G^N$ such that $Eu(x) \ll Eu(x')$.

(iv) Choose any $\bar{x} \in X$ and any balanced subfamily $\mathcal{B}$ of $\mathcal{N} \setminus \{N\}$ with the associated balancing coefficients $\{\lambda_S\}_{S \in \mathcal{B}}$. For each $S \in \mathcal{B}$ choose any $(x^{(S)j})_{j \in S} \in G^S(\bar{x})$, and define T^j-measurable strategy $x^j : T^j \to C^j$ by

$$x^j(\bar{t}^j) := \sum_{S \in \mathcal{B}: S \ni j} \lambda_S \left(x_1^{(S)j}(\bar{t}^j), \sum_{t^{S \setminus \{j\}}} \pi(t^{S \setminus \{j\}} \mid \bar{t}^j) x_2^{(S)j}(\bar{t}^j, t^{S \setminus \{j\}}) \right).$$

Then, $x \in G^N$.

(v) For each $S \in \mathcal{N}$ one of the following two conditions holds true: (1) $G^S(\cdot)$ is a constant correspondence; or (2) For each $j \in S$, there exists a finite subset C_f^j of C^j such that for every $\bar{x} \in X$, $\prod_{j \in S}\{x^j : T^j \to \text{co } C_f^j\} \subset F^S(\bar{x})$, and such that for all $t^j, t'^j \in T^j$ for which $t^j \neq t'^j$, there exist $c^j, c'^j \in C_f^j$ so that $u^j(c^j, t^j) > u^j(c'^j, t^j)$, and $u^j(c'^j, t'^j) > u^j(c^j, t'^j)$. Here, the convex hull of a subset A of a vector space is denoted by co A.

Condition (i.1) is standard in economic theory. Condition (i.2) describes no-externalities, and moreover, imposes a condition which is interpreted in two different ways: First, if each of spaces C_1^j and C_2^j consists only of *pure* (or, *deterministic*) *choices*, then it means the risk-neutrality. Second, if each of spaces C_1^j and C_2^j consists only of *mixed choices* (or *probabilities on pure-choices*) for j, if $u^j(\cdot, t^j)$ is interpreted as the expected utility as a function of j's mixed-choice pairs, and if the underlying von Neumann-Morgenstern utility function is additively separable with respect to the pure choice of the first *interim* period and the pure choice of the second *interim* period, then condition (i.2) is automatically satisfied.

Conditions (ii) and (v.2) are made so that they guarantee existence of a *strictly* Bayesian incentive-compatible strategy bundle; see Abreu and Matsushima's lemma (lemma 4.1.2 of this book).

Condition (iv) is a version of the balancedness condition on the sets, $G^S(\bar{x})$, $S \in \mathcal{N}$, and makes explicit the extent to which the grand coalition has a large feasible-strategy set G^N. It means (1) that for each j the combination of the strategies $\{x^{(S)j}\}_{S \in \mathcal{B}: S \ni j}$ with the convex coefficients $\{\lambda_S\}_{S \in \mathcal{B}: S \ni j}$ is feasible in the grand coalition; and (2) that each member j is insured in the grand coalition to be able to choose $x^j(\bar{t}^j)$ at any state $t \in \{\bar{t}^j\} \times T^{N \setminus \{j\}}$.

Since each C^j is a metric space (condition (i.1)) and T is finite, X^j is also a metric space. Denote by ρ_S the Hausdorff distance on the closed subsets of X^S. The pseudo-metric d on SPACE$_{\text{ne}}$ is defined by:

$$d(\mathcal{S}, \mathcal{S}^\dagger) := \max_{S \in \mathcal{N}} \max_{\bar{x} \in X} \rho_S(G^S(\bar{x}), G^{\dagger S}(\bar{x})).$$

Notice that d may not be a metric, since two distinct sets, $F^S(\bar{x})$ and $F^{\dagger S}(\bar{x})$, may give rise to the identical sets, $G^S(\bar{x}) = G^{\dagger S}(\bar{x})$.

THEOREM 9.1.4 (Ichiishi, Idzik and Zhao, 1994) *Let* (SPACE$_{\text{ne}}$, d) *be the pseudo-metric space of Bayesian societies without externalities satisfying postulates 9.1.1 and 9.1.2, constructed in accordance with assumption 9.1.3. The following is a generic property of a Bayesian society in* (SPACE$_{\text{ne}}$, d): *There exist multitude of Bayesian incentive-compatible strong equilibria, and there exists a Bayesian incentive-compatible strong equilibrium x^* such that x_1^{*j} is 1-1 on T^j.*[2]

We turn to the general case, in which each von Neumann-Morgenstern utility function fully depends on $(x, t) \in X \times T$. We will establish a Bayesian incentive-compatible strong equilibrium generic existence theorem for the general case (theorem 9.1.6), and derive the generic existence theorem for the no-externalities case (theorem 9.1.4) as a corollary.

To establish theorem 9.1.6, one needs to introduce the concepts of *strong* Bayesian incentive compatibility and *weak* Bayesian incentive compatibility, and the associated feasible-strategy sets $\hat{G}_s^S(\bar{x})$ and $\hat{G}_w^S(\bar{x})$. Formally the set $\hat{G}_s^S(\bar{x})$ is defined by: $x^S \in \hat{G}_s^S(\bar{x})$, iff $x^S \in G^S(\bar{x})$, and for all $j \in S$, all $\bar{t}^S \in T^S$, and all $\tilde{t}^j \in T^j$,

$$Eu^j(x^S, \bar{x}^{N \setminus S} \mid \bar{t}^S)$$
$$\geq Eu^j\left((x_1^j(\tilde{t}^j), x_2^j(\tilde{t}^j, \bar{t}^{S \setminus \{j\}}))\right),$$

---

[2]Function x_1^{*j} is T^j-measurable, iff it may be considered a function only of $t^j \in T^j$. A T^j-measurable function is called 1-1 on T^j, if it is 1-1 as a function defined on T^j.

$$(x_1^{S\setminus\{j\}}(\bar{t}^{S\setminus\{j\}}),\ x_2^{S\setminus\{j\}}(\tilde{t}^j,\bar{t}^{S\setminus\{j\}})),\ \bar{x}^{N\setminus S}\ \Big|\ \bar{t}^S\Big).$$

The set $\hat{G}_w^S(\bar{x})$ is defined by: $x^S \in \hat{G}_w^S(\bar{x})$, iff $x^S \in G^S(\bar{x})$, and for all $j \in S$, and all $\bar{t}^j, \tilde{t}^j \in T^j$,

$$Eu^j(x^S, \bar{x}^{N\setminus S} \mid \bar{t}^j)$$
$$\geq\ Eu^j((x_1^j(\tilde{t}^j), x_2^j(\tilde{t}^j,\cdot)), (x_1^{S\setminus\{j\}}, x_2^{S\setminus\{j\}}(\tilde{t}^j,\cdot)), \bar{x}^{N\setminus S} \mid \bar{t}^j).$$

Define the parameterized *ex ante* non-side-payment games $V_{\bar{x}} : \mathcal{N} \to \mathbf{R}^N$, $\bar{x} \in X$, by:

$$V_{\bar{x}}(S) := \left\{ u \in \mathbf{R}^N \ \Big|\ \begin{array}{l} \exists\ x^S \in \hat{G}_w^S(\bar{x}): \\ \forall\ j \in S:\ u_j \leq Eu^j(x^S, \bar{x}^{N\setminus S}) \end{array} \right\} \text{ if } S \neq N;$$

$$V_{\bar{x}}(N) := V(N) = \left\{ u \in \mathbf{R}^N \ \Big|\ \begin{array}{l} \exists\ x^N \in \hat{G}_s^N: \\ \forall\ j \in N:\ u_j \leq Eu^j(x^N) \end{array} \right\}.$$

Assumption 9.1.3 (ii), (iv), (v) are replaced in the general case by the following assumption 9.1.5 (vi), (vii), (viii), respectively.

ASSUMPTION 9.1.5 (vi) There exist a constant strategy for each $j \in N$, $x_m^j : t \mapsto c_m^j$, and a Bayesian incentive compatible strategy bundle for the grand coalition N, $x_M : T \to C$, such that each $x_{M,1}^j$ is T^j-measurable and is 1-1 on T^j, and such that for all S and all $j \in S$, $Eu^j(x_m^S, x_M^{N\setminus S}) < Eu^j(x_M)$.
(vii) For every $\bar{x} \in X$, the non-side-payment game $V_{\bar{x}}$ is balanced, that is, for every balanced subfamily $\mathcal{B}$ of $\mathcal{N}$, $\bigcap_{S \in \mathcal{B}} V_{\bar{x}}(S) \subset V(N)$.
(viii) For each $S \in \mathcal{N}$ one of the following two conditions holds true: (1) $G^S(\cdot)$ is a constant correspondence; or (2) there exists $\hat{x}^S \in X^S$ such that for every $\bar{x} \in X$, $\hat{x}^S \in G^S(\bar{x})$ and for every $j \in S$, every $\bar{t}^S \in T^S$ and every $\tilde{t}^j \in T^j \setminus \{\bar{t}^j\}$, it follows that

$$Eu^j(\hat{x}^S, \bar{x}^{N\setminus S} \mid \bar{t}^S)$$
$$> Eu^j\Big((\hat{x}_1^j(\tilde{t}^j), \hat{x}_2^j(\tilde{t}^j, \bar{t}^{S\setminus\{j\}})),$$
$$(\hat{x}_1^{S\setminus\{j\}}(\bar{t}^{S\setminus\{j\}}), \hat{x}_2^{S\setminus\{j\}}(\tilde{t}^j, \bar{t}^{S\setminus\{j\}})), \bar{x}^{N\setminus S}\ \Big|\ \bar{t}^S\Big).$$

Let $(\{C^j, T^j, u^j, \pi\}_{j \in N}, \{\mathbf{C}_0^S\}_{S \in \mathcal{N}})$ be a list of given data, which satisfies assumptions 9.1.3 (i) and 9.1.5 (vi). The *space of Bayesian societies* SPACE is the set of all Bayesian societies obtained by adding to this list all $\{F^S\}_{S \in \mathcal{N}}$ which satisfy assumptions 9.1.3 (iii), and 9.1.5 (vii) and (viii). The pseudo-metric d on SPACE is defined by:

$$d(\mathcal{S}, \mathcal{S}') := \max_{S \in \mathcal{N}} \max_{\bar{x} \in X} \rho_S(G^S(\bar{x}), G'^S(\bar{x})).$$

THEOREM 9.1.6 (Ichiishi, Idzik and Zhao, 1994) *Let* (SPACE, d) *be the pseudo-metric space of Bayesian societies that satisfy postulates 9.1.1 and 9.1.2, constructed in accordance with assumption 9.1.3 (i) (iii) and assumption 9.1.5. The following is a generic property of a Bayesian society in* (SPACE, d): *There exist multitude of Bayesian incentive compatible strong equilibria, and for at least one of them* x^*, x_1^{*j} *is 1-1 on* T^j *for every* $j \in N$.

Proof of theorem 9.1.6 will be given in the appendix to chapter 9.

Derivation of theorem 9.1.4 from theorem 9.1.6 In view of Abreu and Matsushima's lemma (lemma 4.1.2 of this book), conditions 9.1.3 (ii) and (v) imply conditions 9.1.5 (vi) and (viii), respectively, in the no-externality case. It suffices to show that in the no-externality case, condition 9.1.3 (iv) implies condition 9.1.5 (vii). Let $\mathcal{B}$ be a balanced family with the associated balancing coefficients $\{\lambda_S\}_{S \in \mathcal{B}}$. If $\mathcal{B} \ni N$, there is nothing to prove; so assume $\mathcal{B} \not\ni N$. For each $S \in \mathcal{B}$ there exists $(x^{(S)j})_{j \in S} \in \hat{G}_w^S(\bar{x})$ such that

$$\forall j \in S: \ u_j \leq Eu^j(x^{(S)j}). \tag{9.1}$$

By the definition of $\hat{G}_w^S(\bar{x})$,

$$\forall j \in N: \ (\forall S \in \mathcal{B}: S \ni j): \forall \bar{t}^j, \tilde{t}^j \in T^j:$$

$$Eu^j\left(x^{(S)j} \mid \bar{t}^j\right) \geq Eu^j\left(x_1^{(S)j}(\tilde{t}^j), x_2^{(S)j}(\tilde{t}^j, \cdot) \mid \bar{t}^j\right).$$

That is (using the affine linearity of $u^j(\cdot, \bar{t}^j)$),

$$u^j\left(x_1^{(S)j}(\bar{t}^j), \sum_{t^{S\setminus\{j\}}} \pi(t^{S\setminus\{j\}} \mid \bar{t}^j) x_2^{(S)j}(\bar{t}^j, t^{S\setminus\{j\}}), \bar{t}^j\right)$$

$$\geq u^j\left(x_1^{(S)j}(\tilde{t}^j), \sum_{t^{S\setminus\{j\}}} \pi(t^{S\setminus\{j\}} \mid \bar{t}^j) x_2^{(S)j}(\tilde{t}^j, t^{S\setminus\{j\}}), \bar{t}^j\right). \tag{9.2}$$

Define $x \in G^N(\bar{x})$ as in condition 9.1.3 (iv). Then, by (9.2) for any $\bar{t}^j, \tilde{t}^j \in T^j$,

$$Eu^j(x^j \mid \bar{t}^j)$$

$$= \sum_{S \in \mathcal{B}: S \ni j} \lambda_S u^j\left(x_1^{(S)j}(\bar{t}^j), \sum_{t^{S\setminus\{j\}}} \pi(t^{S\setminus\{j\}} \mid \bar{t}^j) x_2^{(S)j}(\bar{t}^j, t^{S\setminus\{j\}}), \bar{t}^j\right)$$

$$\geq \sum_{S \in \mathcal{B}: S \ni j} \lambda_S u^j\left(x_1^{(S)j}(\tilde{t}^j), \sum_{t^{S\setminus\{j\}}} \pi(t^{S\setminus\{j\}} \mid \bar{t}^j) x_2^{(S)j}(\tilde{t}^j, t^{S\setminus\{j\}}), \bar{t}^j\right)$$

$$= Eu^j(x_1^j(\tilde{t}^j), x_2^j(\tilde{t}^j) \mid \bar{t}^j).$$

Therefore, $x \in \hat{G}_w^N$. Moreover, each x^j depends only upon t^j, so $x \in \hat{G}_s^N$.
Now, by (9.1), for all $j \in N$,

$$\begin{aligned} u_j &\le \sum_{S \in \mathcal{B}: S \ni j} \lambda_S E u^j(x^{(S)j}) \\ &= \sum_{S \in \mathcal{B}: S \ni j} \lambda_S \sum_{\bar{t}^j} \pi(\bar{t}^j) \sum_{t^{S \setminus \{j\}}} \pi(t^{S \setminus \{j\}} \mid \bar{t}^j) E u^j(x^{(S)j} \mid \bar{t}^j) \\ &= \sum_{\bar{t}^j} \pi(\bar{t}^j) E u^j(x^j \mid \bar{t}^j) \\ &= E u^j(x^j) \end{aligned}$$

Thus, $u \in V(N)$. □

9.2 By Contract Execution: A Profit-Center Game with Incomplete Information

A *firm in multidivisional form* (an *M-form firm*, in short) is a corporation in which several divisions (or *profit centers*) are operated semi-autonomously. The significance of M-form firms in the present-day economy has increasingly been recognized since the publication of Chandler's (1962) seminal study of their historical development. Each division in an M-form firm is, to a significant extent, an independent decision-maker.[3] As decision-units of the same corporation, however, these divisions talk to each other and coordinate their production activities. Total profit will then be distributed to the divisions. While coordination of production activities better serves interests of the divisions, there arise conflicts when it comes to the imputation of profit. The divisions therefore agree on a specific plan for coordinated activities/imputation, so that no coalition of divisions can improve upon the plan by its own efforts, that is, the plan is self-enforcing. This is precisely the sort of scenarios that strategic cooperative game theory was developed to analyze.

Radner (1992) formulated the internal organization of an M-form firm as a *profit center game*, a static strategic cooperative game with complete information. He applied the core of the game as the self-enforcing agreements, and studied its properties for several interesting cases.

Ichiishi and Radner (1999) addressed the information revelation process via contract execution within the model of profit center game with incomplete information $\mathcal{D} := (\mathcal{E}_{cp}, p)$, defined as a pair of a specific Bayesian

---
[3] Of course, some decisions are typically made by the firm's central management, *e.g.*, levels and types of capital expenditures, location, and even total numbers of employees.

coalition production economy

$$\mathcal{E}_{cp} := \left(\{\mathbf{R}^{k_m+k_n}, T^j, \text{ profit function}, r^j, \pi\}_{j\in N}, \{Y^j\}_{j\in N}\right)$$

and a price vector of the marketed commodities $p \in \mathbf{R}_+^{k_m}$ (example 2.2.3). Information revelation comes out as part of the planning and implementation process in an M-form firm. Ichiishi and Radner's theory introduces incomplete information into the model of Radner (1992), or equivalently introduces the concrete, economic ingredients of the profit center game into the more abstract formulation of the strategic cooperative game with incomplete information (section 9.1). The approach here is in the spirit of the *managerial theory of the firm*. Because of limitations on the information available to the owners and headquarters of firm, only relatively few parameters of the divisional activities are determined centrally. Then – within those constraints – the divisional managers cooperatively participate in the design and implementation of a corporate plan, including an accounting imputation of the total corporate profit among the divisions.

The two-*interim*-period setup (postulate 9.1.1) and the information revelation process (postulate 9.1.2) are postulated here. Indeed, the two-*interim*-period framework naturally arises from the economic context: The first *interim* period is the setup period, and the second *interim* period is the manufacturing period. The *setup period* is for the divisions' simultaneous decisions about initial investment, setting up their manufacturing processes. The *manufacturing period* is for subsequent decisions about actual choice of an input-output vector, and for imputation of the profit that is made by sale/purchase of the marketed commodities.

In example 2.2.3 we noted the distinction of *marketed commodities* and *nonmarketed commodities*; while a commodity a in the former category has a price p_a established in the market outside the firm, a commodity in the latter category has no price and is used only internally. A *transfer payment problem* addresses determination of prices of nonmarketed commodities according to a nonmarket mechanism. Just as one division can produce an intermediate nonmarketed commodity while some others cannot, production sets, Y^j, $j \in N$, typically differ among the divisions. This usually presupposes *asset specificity* for each division, as emphasized with empirical evidence by Williamson (1975) and Klein, Crawford and Alchian (1978): Some assets are so specialized that the benefit from their use in the firm overweighs their salvage value in the market. Klein, *et al.* (1978) call the difference between the benefit and the salvage value the *quasi rent value*. When the quasi rent value is high, the firm does not bother to sell the assets as scrap, so these resources are classified as nonmarketed commodities. We will focus here on *physical-asset specificity* such as specialized plants designed for production of specific outputs and specialized dies required to produce an intermediate commodity, and the *human-asset specificity* that arises in learning by doing of the human-resource holders.

Differences in production sets also reflects differences in specialization. The degree of specialization may well be private information of division j, and this fact actually motivates the present Bayesian formulation of the M-form firm in the following two ways: First, a feasible net output depends on a type-profile. Second, by interpreting the quantity of a nonmarketed commodity a as

[the actual physical quantity]
× [its efficiency (degree of specialization)],

the amount of a that division j initially holds depends on j's type t^j, hence the notation $r^j(t^j, a)$. [We will see later that the 1-1 property of function r^j plays an essential role in our analysis.]

Due to the specific economic structure of the model, Ichiishi and Radner could establish three *exact* existence theorems for a full-information revealing *ex ante* core plan (Bayesian incentive-compatible strong equilibrium specialized to the present framework), rather than mere *generic* existence theorems of the preceding section (theorem 9.1.4 for space SPACE$_{ne}$ and theorem 9.1.6 for SPACE). Some theorems here are valid even for games that are ruled out from space SPACE.

The first result is under the neoclassical convexity assumption on the production sets (theorem 9.2.5). The assumptions made in theorem 9.2.5 are consistent with the presence of intermediate nonmarketed commodities, so the transfer payment problem is fully answered.

The second result is for a particular class of total production sets that satisfy increasing returns to scale (theorem 9.2.7). Here, a stronger condition than Scarf's (1986) distributiveness is imposed on the total production set Y. Moreover, the nonmarketed commodities are assumed not to be produced as outputs (so intermediate nonmarketed commodities are excluded from the analysis). Roughly speaking, the set Y is required to satisfy: (1) increasing returns to scale with respect to any equiproportional increase in all the type-profile-contingent commodities; (2) strict convexity of the input-requirement sets (in the space of nonmarketed input plans); and (3) strict convexity of the auxiliary concept of production possibility sets (in the space of marketed net output plans) defined in terms of the hypothetical, derivative concept of "cost function." A *non-commodity resource* of a division is a resource owned by or assigned to that division, which is not for sale (so it is not a marketed commodity), and which can not be transferred to another division (so it is not a nonmarketed commodity); e.g., a plant of each division. It should be pointed out that the above properties of Y,

(1) and (3), together implicitly assume that whenever a new net output plan is to be executed, adjustment of non-commodity resources to the new plan is required. In spite of the inability to analyze the role of intermediate nonmarketed commodities in the second result (theorem 9.2.7), the transfer payment problem for exchange of initially held nonmarketed commodities is explicitly solved by a core profit imputation plan.

The third result is for a particular structural relationship among the divisions (theorem 9.2.9). Here, the divisions are divided into the suppliers and the customers. The suppliers produce and supply to the customers nonmarketed intermediate commodities. The customers use these nonmarketed intermediate commodities, produce marketed commodities and bring in profit to the firm. No assumption is made in theorem 9.2.9 on returns to scale.

We will present Ichiishi and Radner's formal analysis now.

Given a type profile $t \in T$, a *profit imputation* of coalition S is a vector $x^S(t) := (x^j(t))_{j \in S}$ whose jth coordinate is the accounting profit attributed to division j. A *profit imputation plan* of coalition S is a function $x^S : T \to \mathbf{R}^S$, $t \mapsto x^S(t)$. Denoting by $y^S : T \to \mathbf{R}^{(k_m + k_n) \cdot \#S}$ a *net output plan*, a pair (x^S, y^S) will henceforth be called a *plan*.

Let K be the index set of all commodities; it is partitioned into the index set K_m of marketed commodities and the index set K_n of nonmarketed commodities. Let K_1 (K_2, resp.) denote the index set for the commodities that are produced/used in the setup period (in the manufacturing period, resp.). The family $\{K_1, K_2\}$ is a partition of K, possibly different from $\{K_n, K_m\}$. Set $k := \#K$, $k_1 := \#K_1$, $k_2 := \#K_2$. A net output plan y^j may be written as

$$y^j = \begin{pmatrix} y_1^j \\ y_2^j \end{pmatrix},$$

where the components of $y_1^j(t)$ (of $y_2^j(t)$, resp.) correspond to K_1 (K_2, resp.). Define y_m^j and y_n^j similarly corresponding to K_m and K_n. Define also $K_{1n} := K_1 \cap K_n$, $k_{1n} := \#K_{1n}$, and define $K_{2n}, K_{1m}, K_{2m}, k_{2n}, k_{1m}$ and k_{2m} similarly. The initial resource function r^j may be written as

$$r^j = \begin{pmatrix} r_1^j \\ r_2^j \end{pmatrix},$$

where r_1^j is a function from T to $\mathbf{R}^{k_{1n}}$, and the components of $r_1^j(t)$ correspond to K_{1n}.

To start the precise description of the scenario, denote by F^S the set of all *technologically attainable* plans of a coalition S, that is, the set of all $\mathcal{T}^S$-measurable functions $(x^S, y^S) : T \to \mathbf{R}^{(1+k) \cdot \#S}$ such that y^S is technologically feasible, i.e.,

$$y^S \in Y^S := \prod_{j \in S} Y^j,$$

and such that the total resource constraint is satisfied within S, i.e.,

$$\forall\, t \in T: \sum_{j \in S} \begin{pmatrix} x^j(t) \\ \mathbf{0} \end{pmatrix} \leq \sum_{j \in S} \begin{pmatrix} p \cdot y_m^j(t) \\ y_n^j(t) + r^j(t) \end{pmatrix}.$$

Notice that negative imputation is allowed.[4]

Suppose coalition S is to form. The members can consider only plans which obey the information pooling rule (postulate 9.1.2); let F'^S to be the set of those *allowable* plans for S, i.e.,

$$F'^S := \left\{ (x^S, y^S) \in F^S \;\middle|\; \begin{array}{l} \forall\, j \in S: \\ y_1^j \text{ is } \mathcal{T}^j\text{-measurable,} \\ (x^j, y_2^j) \text{ is } \hat{\mathcal{T}}^j(y_1^S)\text{- measurable} \end{array} \right\}.$$

The members further restrict their plans to those that satisfy the Bayesian incentive compatibility (postulate 9.A.1 and fact 9.A.2 of the appendix to this chapter). Let $\hat{F}^S$ be the set of allowable, Bayesian incentive-compatible plans for S.

The Bayesian incentive compatibility may be too stringent a postulate that there may not be a strategy in $\hat{F}^N$ which is coalitionally stable. To overcome this difficulty, the headquarters play the role of an insurer. A plan (x^S, y^S) is called *weakly Bayesian incentive-compatible* if for all $j \in S$, and all $\bar{t}^j, \tilde{t}^j \in T^j$, it follows that

$$E(x^j \mid \bar{t}^j) \geq E(x^j \circ (\tilde{t}^j, \text{id}) \mid \bar{t}^j).$$

It is not difficult to show that *for a weakly Bayesian incentive-compatible plan (x^S, y^S), the conditional expectation $E(x^j \mid t^j)$ is independent of t^j, that is, $E(x^j \mid \mathcal{T}^j)$ is a constant function*. This fact motivates the following formulation of the postulate of the *headquarters' insurability*:

POSTULATE 9.2.1 Let (x^N, y^N) be a technologically attainable plan of the grand coalition such that it satisfies the information-revelation process, and $E(x^j \mid \mathcal{T}^j)$ is a constant function for each $j \in N$. Then the plan $((E(x^j \mid \mathcal{T}^j))_{j \in N}, y^N)$ is available to the grand coalition N.

---

[4] In the first main result (theorem 9.2.5), the existence of an equilibrium plan (x^{*N}, y^{*N}), called an *ex ante core plan*, for which $\forall\, t: \forall\, j: x^{*j}(t) \geq 0$ is asserted.

Being an insurer is the only role that the headquarters plays in this game, in addition to participating in the coalitional design of a plan as one of the divisions. By this postulate, division j can receive the accounting profit according to the constant profit imputation plan $E(x^j \mid T^j)$. This is justified if the headquarters is risk-neutral. Moreover, this is an easy task for the headquarters, because it does not have to know the true type of division j (the need for insurance occurs only when $E(x^j \mid t^j)$ is the same for all t^j). This postulate does not reduce the model to a static game. Indeed, while the profit imputation plan for the grand coalition, $(E(x^j \mid T^j))_{j \in N}$, stated in the postulate has a static flavor as a constant function, it is made possible by non-constant net output function y^N, and the latter is subject to the information-revelation process.

Let H^N be the set of all plans (x^N, y^N) for the grand coalition N such that x^N is a constant function, and such that there exists $x'^N : T \to \mathbf{R}^{|N|}$ for which $(x'^N, y^N) \in F'^N$ and $E(x'^j \mid T^j) = x^j$ for every $j \in N$. Notice that every member of H^N is Bayesian incentive-compatible. In the light of the headquarters' insurability, define:

$$\hat{F}^{*S} := \begin{cases} \hat{F}^S, & \text{if } S \neq N, \\ \hat{F}^N \bigcup H^N, & \text{if } S = N. \end{cases}$$

The set $\hat{F}^{*S}$ is the set of all technologically attainable or insurable plans of coalition S that are consistent with the three postulates: the information-revelation process, the Bayesian incentive compatibility, and the headquarters' insurability. Plan (x^S, y^S) is a candidate for coalition S's agreement, iff $(x^S, y^S) \in \hat{F}^{*S}$.

DEFINITION 9.2.2 An *ex ante core plan* of a profit-center game with incomplete information $\mathcal{D}$ is a Bayesian incentive-compatible strong equilibrium: It is a plan (x^{*N}, y^{*N}) of the grand coalition N such that
(i) $(x^{*N}, y^{*N}) \in \hat{F}^{*N}$, and
(ii) it is not true that

$$\exists\, S \in \mathcal{N} : \exists\, (x^S, y^S) \in \hat{F}^{*S} : \forall\, j \in S : Ex^j > Ex^{*j}.$$

A core plan (x^{*N}, y^{*N}) is called *full-information revealing*, if for every $j \in N$, y_1^{*j} is 1-1 on T^j. In this case, the updated algebra $\hat{T}^j(y_1^{*N})$ becomes the full communication system 2^T. The main results of this section are three existence theorems for a full-information revealing core plan of a profit-center game with incomplete information $\mathcal{D}$.

There are two sets of basic assumptions made in all the three theorems. The first set (assumption 9.2.3 below) imposes conditions on the production set of each division. In particular, (iii) says that zero production activity is possible; (iv) means free disposal; and (v) means the impossibility of the Land of Cockaigne, that is, only a finite quantity of outputs is produced from a finite quantity of inputs.

ASSUMPTION 9.2.3 (Basic Assumptions on the Production Sets)
For each coalition S, its total production set $Y(S)$ is given as
(i) $Y(S) := \sum_{j \in S} Y^j$.
For each j,
(ii) the production set Y^j is closed in $\mathbf{R}^{k|T|}$;
(iii) $\mathbf{0} \in Y^j$;
(iv) $Y^j - \mathbf{R}_+^{k|T|} \subset Y^j$;
(v) for each $y_n^j \in \mathbf{R}^{k_n|T|}$, the production possibility set $\{y_m^j \in \mathbf{R}^{k_m|T|} \mid (y_m^j, y_n^j) \in Y^j\}$ is bounded from above.

We make two comments on the production sets. First, to take the individual production sets Y^j as given data, and derive the coalitional total production sets as in assumption 9.2.3 (i) presupposes that there are no external economies or diseconomies. On the other hand, one may adopt a more general approach in which coalitional total production sets $Y(S)$, $S \in \mathcal{N}$, are given data; in this case, external economies can be formulated as $Y(S) \supset \sum_{j \in S} Y(\{j\})$. Although the main results (theorems 9.2.5, 9.2.7 and 9.2.9) are established for the non-externality case, generalizations to the externality case turn out to be straightforward.

Second, a specific instance of a production set is given by

$$Y^j = \prod_{t \in T} Y^j(t), \ Y^j(t) \subset \mathbf{R}^k.$$

Here, $Y^j(t)$ is a *t-contingent production set* of division j. In this case, given type-profile $\bar{t}$, $\{\bar{y}_m^j(t)\}_{t \in T \setminus \bar{t}}$ and $\bar{y}_n^j$, the production possibility set for marketed commodities at $\bar{t}$,

$$\{y_m^j(\bar{t}) \in \mathbf{R}^{k_m} \mid (y_m^j(\bar{t}), \{\bar{y}_m^j(t)\}_{t \in T \setminus \bar{t}}, \bar{y}_n^j) \in Y^j\},$$

is determined only by $\bar{y}_n^j(\bar{t})$, and is independent of net outputs $\bar{y}^j(t)$'s chosen at the other type-profiles $t \neq \bar{t}$. The present general formulation, on the other hand, allows for the possibility that the production possibility set for marketed commodities at $\bar{t}$ depends on the entire menu of non-marketed net outputs across all type-profiles, $\{\bar{y}_n^j(t)\}_{t \in T}$, as well as on $\{\bar{y}_m^j(t)\}_{t \neq \bar{t}}$. The need for this general formulation arises if, for the production of net marketed output plan y_m^j, an adjustment of specific effort and/or non-commodity resources is required in addition to the choice of a net non-marketed output plan y_n^j, and different adjustments of effort/resources are required for different y_m^j's. This dependence (that is, the dependence of the production possibility set for marketed commodities at each t on the entire nonmarketed net output plan y_n^j) turns out to be crucial for the existence result in the case of increasing returns to scale (theorem 9.2.7).

The second set of basic assumptions (assumption 9.2.4 below) imposes conditions on the resource function of each division. In particular, (i) says that there are nonmarketed commodities which are used or produced in the setup period. Assumption (ii) says that r_1^j takes different values for different types of division j; the fifth through the sixth paragraphs of the present section have given a detailed account of the necessity of this 1-1 property, when i's type is defined as a state of its technology embodied in its initial investment. Assumption (iii) says that all resources are nonnegative.

ASSUMPTION 9.2.4 (Basic Assumptions on the Resource Functions) (i) $K_{1n} \neq \emptyset$;
For each $j \in N$,
(ii) the function r_1^j is 1-1 on T^j;
(iii) $r^j(t^j) \geq \mathbf{0}$, for all $t^j \in T^j$.

The first result is the following existence theorem for the neoclassical case of convex technology:

THEOREM 9.2.5 (Ichiishi and Radner, 1999) *Let $\mathcal{D}$ be a profit-center game with incomplete information which satisfies the three postulates: the information-revelation process (postulate 9.1.2), the Bayesian incentive compatibility (postulate 9.A.1) and the headquarters' insurability (postulate 9.2.1), and the two basic assumptions, 9.2.3 and 9.2.4. Assume, moreover, for each $j \in N$ that the ex ante probability π is the product probability of π^j, $j \in N$, where π^j is a probability on T^j, and that the production set Y^j is convex for each $j \in N$. Then there exists a full-information revealing ex ante core plan of the game.*

The second result is an existence theorem for a particular technology which satisfies increasing returns to scale. The theorem, however, excludes nonmarketed intermediate commodities. Nevertheless, nonmarketed commodities held as the initial resources can be exchanged among the divisions, and a core imputation plan describes in part transfer payments for these exchanges.

Introduce a fictitious price vector, $q := (q(t))_{t \in T}$, $q(t) = (q_1(t), q_2(t)) \in \mathbf{R}_+^{k_{1n}} \times \mathbf{R}_+^{k_{2n}}$, for type-profile-contingent nonmarketed commodities; this is to utilize a cost function concept in the analysis. Set $q(t) := (q(t,a))_{a \in K_n}$.

One may choose the unit simplex,

$$Q := \left\{ q : T \to \mathbf{R}^{k_n}_+ \;\middle|\; \sum_{t \in T} \sum_{a \in K_n} q(t,a) = 1 \right\},$$

as the price domain.

The exclusion of nonmarketed intermediate commodities means that all the nonmarketed commodities are only used as inputs. Given a marketed net output plan $y_m : T \to \mathbf{R}^{k_m}$, therefore, one may define the *input-requirement set* as

$$S_{YN}(y_m) := \left\{ -y_n : T \to \mathbf{R}^{k_n} \;\middle|\; \begin{array}{l} (y_m, y_n) \leq \sum_{j \in N}(y_m^j, y_n^j), \\ \forall\, j \in N : (y_m^j, y_n^j) \in Y^j, \\ y_1^j \text{ is } \mathcal{T}^j\text{-measurable} \end{array} \right\}.$$

A cost function is then defined as the function $g : Q \times \mathbf{R}^{k_m|T|} \to \mathbf{R}_+$ which associates to each pair of a fictitious price vector and a marketed net output plan (q, y_m) the minimal cost of using nonmarketed commodities in order to produce y_m under q. Formally, it is defined by

$$g(q, y_m) := \inf \left\{ -\sum_{t \in T} q(t) \cdot y_n(t) \;\middle|\; -y_n \in S_{YN}(y_m) \right\}.$$

The concept of weak Bayesian incentive compatibility was introduced earlier (in the paragraph preceding postulate 9.2.1). Another type of cost function which accommodates the weak Bayesian incentive compatibility, $\hat{c}_w^S : Q \times \overset{\circ}{\mathbf{R}}_+ \to \mathbf{R}_+$, is now defined: In order to do so, choose any profit level η, and consider the set of all pairs of a profit imputation plan and a total net output plan $(x^S, \sum_{j \in S} y^j)$ such that (1) each y^j is technologically feasible, (2) the total imputation $\sum_{j \in S} x^j(t)$ is met by the total profit $\sum_{j \in S} p \cdot y_m^j(t)$ for every possible type-profile t, and (3) the conditional expectations $E(x^j \mid t^j)$ of the imputations x^j given j's private information $\mathcal{T}^j$, $j \in S$, are summed up at least to η for any possible type-profile t. Formally, the set is given as:

$$C^S(\eta) := \left\{ \left(x^S, \sum_{j \in S} y^j \right) \;\middle|\; \begin{array}{l} y^j \in Y^j, \\ \forall\, t \in T : \sum_{j \in S} x^j(t) \leq \sum_{j \in S} p \cdot y_m^j(t), \\ \eta \leq \sum_{j \in S} \min_{t \in T} E(x^j \mid \mathcal{T}^j)(t), \\ (x^S, y_m^S) \text{ is } \mathcal{T}^S\text{-measurable}, \\ \forall\, j \in S : y_1^j \text{ is } \mathcal{T}^j\text{-measurable} \end{array} \right\}.$$

Define the mathematical programming problem,

$$\text{Problem } P^S(q, \eta) : \quad \text{Minimize} \quad -\sum_{t \in T} q(t) \cdot y_n(t),$$

9: Information Revelation

$$\text{subject to} \quad (x^S, y) \in C^S(\eta).$$

The required cost function is the optimal value of problem $P^S(q, \eta)$:

$$\hat{c}_w^S(q, \eta) := \inf \left\{ -\sum_{t \in T} q(t) \cdot y_n(t) \,\middle|\, (x^S, y) \in C^S(\eta) \right\}.$$

For a net output plan y^j for which y_1^j is T^j-measurable, the cost is

$$-\sum_{t \in T} q(t) \cdot y_n^j(t) = -\sum_{t^j \in T^j} \left\{ \sum_{t^{N\setminus\{j\}} \in T^{N\setminus\{j\}}} q_1(t^j, t^{N\setminus\{j\}}) \right\} \cdot y_{1n}^j(t^j)$$
$$- \sum_{t \in T} q_2(t) \cdot y_{2n}^j(t),$$

so problem $P^S(q, \eta)$ has an optimal solution, if

$$\forall j \in S: \forall t^j \in T^j: \sum_{t^{N\setminus\{j\}} \in T^{N\setminus\{j\}}} q_1(t^j, t^{N\setminus\{j\}}) \gg 0,$$

and

$$\forall t \in T: q_2(t) \gg 0.$$

The same remark applies to the earlier minimization problem that defines $g(q, y_m)$. This fact motivates the following definitions:

$$Q_+ := \bigcap_{j \in N} \bigcap_{t^j \in T^j} \left\{ q \in Q \,\middle|\, \sum_{t^{N\setminus\{j\}} \in T^{N\setminus\{j\}}} q_1(t^j, t^{N\setminus\{j\}}) \gg 0 \right\}$$
$$\bigcap_{t \in T} \{q \in Q \,|\, q_2(t) \gg 0\},$$
$$Q_0 := Q \setminus Q_+.$$

Problem $P^S(q, \eta)$ and the problem for $g(q, y_m)$ have optimal solutions, if $q \in Q_+$.

The second theorem of this section (theorem 9.2.7) is based on the following assumption 9.2.6, which replaces the convexity assumption made in theorem 9.2.5. It is a strengthened version of Scarf's (1986) distributiveness assumption. Condition (i) of assumption 9.2.6 means increasing returns to scale. Condition (ii) means that nonmarketed commodities are only used as inputs and cannot be produced; in particular it excludes nonmarketed intermediate commodities. Condition (iii) says that $\hat{c}_w^N$ can be continuously extended from $Q_+ \times \overset{\circ}{\mathbf{R}}_+$ to $Q \times \overset{\circ}{\mathbf{R}}_+$. Condition (iv) means the convexity of each production possibility set. Condition (v) means diminishing marginal rate of technical substitution.

ASSUMPTION 9.2.6 (Strongly Distributive Total Production Set of the Grand Coalition)
(i) For any $y^N \in Y^N$ and any real number $\alpha > 1$, it follows that $\alpha y^N \in Y^N$;
(ii) for any $j \in N$, $Y^j \subset \mathbf{R}^{k_m|T|} \times (-\mathbf{R}_+^{k_n|T|})$;
(iii) cost function $\hat{c}_w^S : Q \times \overset{\circ}{\mathbf{R}}_+ \to \mathbf{R}_+$ is continuous;
(iv) for any $q \in Q_+$, the cost function $g(q, \cdot)$ is strictly quasi-convex;
(v) for each y_m, the input-requirement set $S_{Y^N}(y_m)$ is strictly convex.

A stronger form of exclusion of a nonmarketed intermediate commodity is assumed in theorem 9.2.7: Condition (i) in theorem 9.2.7 says that if all nonmarketed commodities are used as inputs at any type-profile, then a positive profit can be made at any type-profile; in particular, it excludes those divisions that produce only nonmarketed intermediate commodities (the firm with such a division will be analyzed in the third existence result). Condition (ii) says that each division is endowed with all nonmarketed commodities at the outset at any type-profile, which again excludes a nonmarketed intermediate commodity.

THEOREM 9.2.7 (Ichiishi and Radner, 1999) *Let $\mathcal{D}$ be a profit-center game with incomplete information which satisfies the three postulates: the information-revelation process (postulate 9.1.2), the Bayesian incentive compatibility (postulate 9.A.1) and the headquarters' insurability (postulate 9.2.1), and the two basic assumptions, 9.2.3 and 9.2.4, and assumption 9.2.6. Assume moreover that*
(i) *for any $y_n^j \in \mathbf{R}^{k_n|T|}$ for which $y_n^j \ll \mathbf{0}$, there exists $y'^j \in Y^j$ such that $y_n^j \leq y_n'^j$, $y_1'^j$ is $\mathcal{T}^j$-measurable, and $p \cdot y_m'^j(t) > 0$ for all $t \in T$; and*
(ii) *$r^j(t^j) \gg \mathbf{0}$, for every $j \in N$ and every $t \in T$.*
Then there exists a full-information revealing core plan of the game.

Proof of theorem 9.2.7 will be given in the appendix to this chapter.

The preceding two existence results focused on the role of returns to scale, non-increasing or increasing. The third result, on the other hand, focuses on the role of a structural relationship between two types of divisions, the suppliers and the customers. The *suppliers* produce and supply to the customers nonmarketed intermediate commodities. The *customers* use these nonmarketed intermediate commodities, produce marketed commodities and bring in profit to the firm. The products of all suppliers are needed for each customer's production activities, hence the term *complementary* suppliers. No assumption on returns to scale is made. The idea about this relationship goes back to Radner (1992, subsection 7.2). The following are a formal treatment of its simplified version in the context of incomplete information.

The division set, N, is partitioned into the supplier set, N_s, and the customer set, N_c. Let K_{n_s} be the set of nonmarketed intermediate commodities, a subset of K_n. A net output plan y^j is denoted by $(y^j_m, y^j_{n_s}, y^j_{n_z})$, where the subvectors $y^j_{n_s}$ and $y^j_{n_z}$ correspond to K_{n_s} and $K_n \setminus K_{n_s}$, respectively.

The following assumption 9.2.8 highlights the role of complementary suppliers of nonmarketed intermediate commodities. In particular, (i) says that a supplier can produce only nonmarket intermediate commodities; (ii) says that all the suppliers are needed in order to produce all nonmarketed intermediate commodities; (iii) says that for a customer to produce a marketed commodity, all the nonmarketed intermediate commodities are needed; (iv) says that a coalition of a customer and all the suppliers can make a positive profit if all the nonmarketed non-intermediate commodity resources are available (even if in arbitrarily small amount); (v.1) says that no nonmarketed intermediate commodities are available as initial resources (so that they have to be produced by suppliers); and (v.2) says that all the nonmarketed non-intermediate commodities are available as initial resources in the grand coalition.

ASSUMPTION 9.2.8 ([Complementary Supplier]-Customer Relationship)
(i) For each supplier $j \in N_s$, if $y^j \in Y^j$, then $y^j_m \leq \mathbf{0}$ and $y^j_{n_z} \leq \mathbf{0}$.
(ii) If $y^j \in Y^j$ for each supplier $j \in N_s$ and if $\sum_{j \in N_s} y^j_{n_s}(t) \gg \mathbf{0}$ for some $t \in T$, then for this t, $y^j_{n_s}(t) > \mathbf{0}$ for every $j \in N_s$.
(iii) For each customer $j \in N_c$ and each $t \in T$, if $[y^j \in Y^j$ and $\neg\, y^j_{n_s}(t) \ll \mathbf{0}]$, then $y^j_m(t) \leq \mathbf{0}$.
(iv) For any $j \in N_c$ and any $\varepsilon > 0$, there exists $y^{N_s \cup \{j\}} \in Y^{N_s \cup \{j\}}$ such that each y^i_1 is $\mathcal{T}^i$-measurable, $i \in N_s \cup \{j\}$, and

$$\forall\, (t, a) \in T \times K_{n_c} : \quad \sum_{i \in N_s \cup \{j\}} y^i_{n_s}(t, a) > -\varepsilon,$$
$$\forall\, t \in T : \quad p \cdot \sum_{i \in N_s \cup \{j\}} y^i_m(t) > 0.$$

(v)(1) For each division $j \in N$, $r^j_{n_s} = \mathbf{0}\ (\in \mathbf{R}^{k_{n_s}|T|})$; (2) $\sum_{j \in N} r^j_{n_z} \gg \mathbf{0}$.

A customer has to use some nonmarketed non-intermediate commodities in the setup period, *e.g.*, human asset. Therefore, assumption 9.2.8 is consistent with the present setup of supplier-customer relationship. The following existence theorem is valid regardless of the nature of returns to scale.

THEOREM 9.2.9 (Ichiishi and Radner, 1999) *Let $\mathcal{D}$ be a profit-center game with incomplete information which satisfies the three postulates: the information-revelation process (postulate 9.1.2), the Bayesian* *incentive compatibility (postulate 9.A.1) and the headquarters' insurability (postulate 9.2.1), and the two basic assumptions, 9.2.3 and 9.2.4, and assumption 9.2.8. Then there exists a full-information revealing core plan of the game.*

Before providing proofs of theorems 9.2.5 and 9.2.9, we present a lemma: In this lemma, $y_1^j = (y_{1m}^j, y_{1n}^j)$, so vector $y_{1n}^j(t^j)$ is a nonmarketed commodity bundle used as inputs in the setup period, whose components are measured by negative real numbers according to the usual sign convention on inputs and outputs.

LEMMA 9.2.10 (Ichiishi and Radner, 1999) *Suppose that for each j there exists a function $y_{1n}^j : T^j \to \mathbf{R}^{k_{1n}}$ such that*

$$\forall\, t \in T : \quad -\sum_{j \in N} y_{1n}^j(t^j) \leq \sum_{j \in N} r_1^j(t^j). \tag{9.3}$$

Then, for each j there exists a function $\overline{\overline{y}}_{1n}^j : T^j \to \mathbf{R}^{k_{1n}}$ such that $\overline{\overline{y}}_{1n}^j \leq y_{1n}^j$ and

$$\forall\, t \in T : \quad -\sum_{j \in N} \overline{\overline{y}}_{1n}^j(t^j) = \sum_{j \in N} r_1^j(t^j).$$

Proof Step 1. For commodity a, the ath component of $y^j(t^j)$ is denoted by $y^j(t^j, a)$. It will be shown first that if

$$\exists\, \bar{t} \in T : \ \exists\, \hat{a} \in K_{1n} : \ -\sum_{j \in N} y^j(\bar{t}^j, \hat{a}) < \sum_{j \in N} r^j(\bar{t}^j, \hat{a}), \tag{9.4}$$

then

$$\exists\, j : \forall\, t^{N \setminus \{j\}} : -y^j(\bar{t}^j, \hat{a}) - \sum_{i \in N \setminus \{j\}} y^i(t^i, \hat{a}) < r^j(\bar{t}^j, \hat{a}) + \sum_{i \in N \setminus \{j\}} r^i(t^i, \hat{a}).$$

This will be done in the following steps 2-4. Only in these steps 2-4, simplifying notation will be used, so that $y^j(t^j, \hat{a})$, $y^j(\bar{t}^j, \hat{a})$, $y^j(t'^j, \hat{a})$, $r^j(t^j, \hat{a})$, $r^j(\bar{t}^j, \hat{a})$, $r^j(t'^j, \hat{a})$, etc., will be denoted by n^j, $\bar{n}^j$, n'^j, r^j, $\bar{r}^j$, r'^j, etc., respectively, and the set of divisions N will be identified with the set of integers $\{1, 2, \cdots, N\}$ (where the last integer in the set is $|N|$, by abuse of notation). Condition (9.4) is then re-written as:

$$\exists\, \bar{t} : -\bar{n}^1 - \bar{n}^2 - \bar{n}^3 - \cdots - \bar{n}^N < \bar{r}^1 + \bar{r}^2 + \bar{r}^3 + \cdots + \bar{r}^N.$$

Step 2. If for all $(t^2, t^3, \cdots, t^N)$,

$$-\bar{n}^1 - n^2 - n^3 - \cdots - n^N < \bar{r}^1 + r^2 + r^3 + \cdots + r^N,$$

9: Information Revelation

then there is nothing to prove. So assume that there exists $(t'^2, t'^3, \cdots, t'^N)$ such that

$$-\bar{n}^1 - n'^2 - n'^3 - \cdots - n'^N = \bar{r}^1 + r'^2 + r'^3 + \cdots + r'^N.$$

Then, for all t^1,

$$-n^1 - \bar{n}^2 - \bar{n}^3 - \cdots - \bar{n}^N \;<\; r^1 + \bar{r}^2 + \bar{r}^3 + \cdots + \bar{r}^N. \qquad (9.5)$$

Indeed, if there exists $\tilde{t}^1$ for which

$$-\tilde{n}^1 - \bar{n}^2 - \bar{n}^3 - \cdots - \bar{n}^N = \tilde{r}^1 + \bar{r}^2 + \bar{r}^3 + \cdots + \bar{r}^N,$$

then by adding the two equalities,

$$\begin{aligned}
& (-\bar{n}^1 - \bar{n}^2 - \bar{n}^3 - \cdots - \bar{n}^N) + (-\tilde{n}^1 - n'^2 - n'^3 - \cdots - n'^N) \\
&= (-\bar{n}^1 - n'^2 - n'^3 - \cdots - n'^N) + (-\tilde{n}^1 - \bar{n}^2 - \bar{n}^3 - \cdots - \bar{n}^N) \\
&= (\bar{r}^1 + r'^2 + r'^3 + \cdots + r'^N) + (\tilde{r}^1 + \bar{r}^2 + \bar{r}^3 + \cdots + \bar{r}^N) \\
&= (\bar{r}^1 + \bar{r}^2 + \bar{r}^3 + \cdots + \bar{r}^N) + (\tilde{r}_1 + r'^2 + r'^3 + \cdots + r'^N).
\end{aligned}$$

By (9.4),

$$-\tilde{n}^1 - n'^2 - n'^3 - \cdots - n'^N > \tilde{r}^1 + r'^2 + r'^3 + \cdots + r'^N,$$

which contradicts (9.3).

Step 3. If for all $(t^1, t^3, t^4, \cdots, t^N)$,

$$-n^1 - \bar{n}^2 - n^3 - \cdots - n^N < r^1 + \bar{r}^2 + r^3 + \cdots + r^N,$$

then there is nothing to prove. So assume that there exists $(t''^1, t''^3, t''^4, \cdots, t''^N)$ such that

$$-n''^1 - \bar{n}^2 - n''^3 - \cdots - n''^N = r''^1 + \bar{r}^2 + r''^3 + \cdots + r''^N.$$

Then, for all (t^1, t^2),

$$-n^1 - n^2 - \bar{n}^3 - \cdots - \bar{n}^N \;<\; r^1 + r^2 + \bar{r}^3 + \cdots + \bar{r}^N. \qquad (9.6)$$

Indeed, if there exists $(\tilde{t}^1, \tilde{t}^2)$ for which

$$-\tilde{n}^1 - \tilde{n}^2 - \bar{n}^3 - \cdots - \bar{n}^N = \tilde{r}^1 + \tilde{r}^2 + \bar{r}^3 + \cdots + \bar{r}^N,$$

then by adding the two equalities,

$$\begin{aligned}
& (-n''^1 - \bar{n}^2 - \bar{n}^3 - \cdots - \bar{n}^N) + (-\tilde{n}^1 - \tilde{n}^2 - n''^3 - \cdots - n''^N) \\
&= (r''^1 + \bar{r}^2 + \bar{r}^3 + \cdots + \bar{r}^N) + (\tilde{r}^1 + \tilde{r}^2 + r''^3 + \cdots + r''^N).
\end{aligned}$$

By (9.5),
$$-\tilde{n}^1 - \tilde{n}^2 - n''^3 - \cdots - n''^N > \tilde{r}^1 + \tilde{r}^2 + r''^3 + \cdots + r''^N,$$
which contradicts (9.3).

Step 4. Continuing this way, one can prove that if for each i, $1 \leq i \leq N-1$, there exists $(t^{i,1}, \cdots, t^{i,i-1}, t^{i,i+1}, \cdots, t^{i,N})$ for which
$$-n^{i,1} - n^{i,2} - \cdots - n^{i,i-1} - \bar{n}^i - n^{i,i+1} - n^{i,i+2} - \cdots - n^{i,N}$$
$$= r^{i,1} + r^{i,2} + \cdots + r^{i,i-1} + \bar{r}^i + r^{i,i+1} + r^{i,i+2} + \cdots + r^{i,N},$$
then for all $(t^1, t^2, \cdots, t^{N-1})$,
$$-n^1 - n^2 - \cdots - n^{N-1} - \bar{n}^N < r^1 + r^2 + \cdots + r^{N-1} + \bar{r}^N.$$
This is precisely the required result stated in step 1.

Step 5. Suppose (9.4) holds true. Then, by step 1, one may choose j_1 and $\varepsilon > 0$ such that for all $t^{N \setminus \{j_1\}}$
$$-[n^{j_1}(\bar{t}^{j_1}, \hat{a}) - \varepsilon] - \sum_{i \in N \setminus \{j_1\}} n^i(t^i, \hat{a}) \leq r^{j_1}(\bar{t}^{j_1}, \hat{a}) + \sum_{i \in N \setminus \{j_1\}} r^i(t^i, \hat{a}),$$
with equality for at least one $t^{N \setminus \{j_1\}}$. Define $n_1^{(1)i} : T^i \to \mathbf{R}^{k_{1n}}$ by
$$\forall\, a \in K_{1n} : n^{(1)i}(t^i, a) := \begin{cases} n^{j_1}(\bar{t}^{j_1}, \hat{a}) - \varepsilon, & \text{if } (i, t^i, a) = (j_1, \bar{t}^{j_1}, \hat{a}); \\ n^i(t^i, a), & \text{otherwise.} \end{cases}$$

The function $n_1^{(1)N}$ satisfies (9.3), and $n_1^{(1)i} \leq n_1^i$. Moreover, by the choice of ε, the number of strict inequalities in (9.3) is less for $n_1^{(1)N}$ than for n_1^N. If (9.4) is true for $n_1^{(1)N}$, repeat the procedure (steps 1-4) and obtain $n_1^{(2)N}$ ($< n_1^{(1)N}$). At each repetition, the number of strict inequalities in (9.3) strictly decreases. After finitely many steps, one obtains equality in (9.3). □

We have introduced the concept of weak incentive compatibility earlier (in the paragraph preceding postulate 9.2.1). Define
$$G^S := \{(x^S, y^S) \in F^S \mid \forall\, j \in S : y_1^j \text{ is } \mathcal{T}^j\text{-measurable.}\},$$
$$\hat{G}_w^S := \left\{ (x^S, y^S) \in G^S \,\middle|\, \begin{array}{l} (x^S, y^S) \text{ is weakly Bayesian} \\ \text{incentive-compatible.} \end{array} \right\},$$
and define the non-side-payment games $\hat{V}_w : \mathcal{N} \to \mathbf{R}^N$ and $\hat{V}^* : \mathcal{N} \to \mathbf{R}^N$ by
$$\hat{V}_w(S) := \{u \in \mathbf{R}^N \mid \exists\, (x^S, y^S) \in \hat{G}_w^S : \forall\, j \in S : u_j \leq Ex^j\},$$
$$\hat{V}^*(S) := \{u \in \mathbf{R}^N \mid \exists\, (x^S, y^S) \in \hat{F}^{*S} : \forall\, j \in S : u_j \leq Ex^j\}.$$

Proof of Theorem 9.2.5 The non-side-payment game $\hat{V}_w$ is balanced. Indeed, let $\mathcal{B}$ be a balanced subfamily of $\mathcal{N}$ with the associated balancing coefficients $\{\lambda_S\}_{S\in\mathcal{B}}$, and choose any $u \in \bigcap_{S\in\mathcal{B}} \hat{V}_w(S)$. For each $S \in \mathcal{B}$, there exists $(x^{(S)j}, y^{(S)j})_{j\in S} \in \hat{G}_w^S$ such that $u_j \leq Ex^{(S)j}$ for every $j \in S$. Define (x^N, y^N) by

$$(x^j, y^j) := \sum_{S\in\mathcal{B}: S \ni j} \lambda_S (x^{(S)j}, y^{(S)j}).$$

By the present convexity condition on Y^j, $(x^N, y^N) \in G^N$. Since each $E(x^{(S)j} \mid \mathcal{T}^j)$ is a constant function on T by the weak incentive compatibility, so is $E(x^j \mid \mathcal{T}^j)$ as a weighted average of these functions. Therefore, $(x^N, y^N) \in \hat{G}_w^N$. Since $u_i \leq Ex^i$, it follows that $u \in \hat{V}_w(N)$.

By Scarf's theorem for nonemptiness of the core (theorem 8.A.2 of this book), one can choose u^* in the core of game $\hat{V}_w$. Let $(\bar{x}^N, \bar{y}^N) \in \hat{G}_w^N$ be a plan that gives rise to u^*. Clearly, $\hat{G}_w^S \supset \hat{F}^S$, so u^* cannot be improved upon by any proper coalitions in game $\hat{V}^*$.

By lemma 9.2.10, for each j there exists a $\mathcal{T}^j$-measurable function $\bar{\bar{y}}_{1n}^j : T \to \mathbf{R}^{k_{1n}}$ such that

$$\bar{\bar{y}}_{1n}^j \leq \bar{y}_{1n}^j,$$

and

$$-\sum_{j\in N} \bar{\bar{y}}_{1n}^j = \sum_{j\in N} r_1^j.$$

Each $\bar{\bar{y}}_{1n}^j$ is 1-1 on $\mathcal{T}^j$. Indeed, this follows from the identity,

$$\forall t \in T : \bar{\bar{y}}_{1n}^j (t^j) + r_1^j(t^j) = -\sum_{i\in N\setminus\{j\}} (\bar{\bar{y}}_{1n}^i (t^i) + r_1^i(t^i)),$$

and the facts that the right-hand side is constant once $t^{N\setminus\{j\}}$ is fixed, and that $r_1^j(\cdot)$ is 1-1 on $\mathcal{T}^j$ (basic assumption 9.2.4 (ii)).

Define $\bar{\bar{x}}^j := E(\bar{x}^j \mid \mathcal{T}^j)$, and define the plan (x^{*N}, y^{*N}) by

$$x^{*N} := \bar{\bar{x}}^N, \quad y_{1m}^{*N} := \bar{y}_{1m}^N, \quad y_{1n}^{*N} := \bar{\bar{y}}_{1n}^N, \quad y_2^{*N} := \bar{y}_2^N.$$

By the free disposal assumption (assumption 9.2.3 (iv)), $(\bar{x}^N, y^{*N}) \in \hat{G}_w^N$. Since y_{1n}^{*j} is 1-1 on $\mathcal{T}^j$ for all j, it follows that $\hat{\mathcal{T}}^j(y_1^{*N})$ is the finest algebra 2^T for every j, so $(\bar{x}^N, y^{*N}) \in F'^N$. By the headquarters' insurability (postulate 9.2.1), $(x^{*N}, y^{*N}) \in \hat{F}^{*N}$. Notice that $E\bar{\bar{x}}^j = E\bar{x}^j \geq u_j^*$. It is also easy to check that u^* lies in the Pareto frontier of $\hat{V}^*(N)$. Thus, u^* is in the core of game $\hat{V}^*$. The plan (x^{*N}, y^{*N}) is full-information revealing. □

Proof of theorem 9.2.9 is a straightforward consequence of the following lemma 9.2.11 (applied to game $\hat{V}_w$) and the argument in the last two paragraphs of the proof of theorem 9.2.5, so the details are left to the reader. Let $\partial V(N)$ be the Pareto frontier of set $V(N)$. Assumption (iii) of lemma 9.2.11 means that $\partial V(N)$ is strictly negatively sloped.

LEMMA 9.2.11 *Let $V : \mathcal{N} \to \mathbf{R}^N$ be a non-side-payment game. Assume:*
(i) *V is superadditive, i.e., for any disjoint coalitions S_1 and S_2 it follows that $V(S_1) \cap V(S_2) \subset V(S_1 \cup S_2)$.*
(ii) *There exist a partition $\{N_s, N_c\}$ of N, and $\underline{u} \in \mathbf{R}^N$ such that if $S \not\supset N_s$, then $V(S) = [\{\underline{u}^S\} - \mathbf{R}_+^S] \times \mathbf{R}^{N \setminus S}$.*
(iii) *For any $\varepsilon > 0$ and any $u \in \partial V(N)$ for which $u > \underline{u}$, there exist $u' \in V(N)$ and $j \in \{i \in N \mid u_i > \underline{u}_i\}$, such that $u'_j = u_j - \varepsilon$ and $u'_i > u_i$ for all $i \in N \setminus \{j\}$.*
Choose any $u^* \in \partial V(N)$ such that

$$u_j^* \geq \underline{u}_j \quad \text{for all } j \in N_s, \quad \text{and}$$
$$u_j^* = \underline{u}_j \quad \text{for all } j \in N_c.$$

Then u^* is in the core of V.

Proof Suppose there exists $S \in \mathcal{N}$ such that $u^* \in \overset{\circ}{V}(S)$. Then

$$N_s \subset S \neq N.$$

Let $u' \in V(S)$ be such that

$$u'_j > u_j^* \quad \text{for all } j \in S,$$
$$u'_j = \underline{u}_j \quad \text{for all } j \in N \setminus S \, (\subset N_c).$$

By the superadditivity, $u' \in V(N)$. By the strict negative-slopedness of $\partial V(N)$, there exists $u'' \in V(N)$ such that $u'' \gg u^*$, which contradicts the choice of u^* as a point of $\partial V(N)$. □

Ichiishi and Sertel (1998) continued study of the profit-center game with incomplete information. They first pointed out that, contrary to *ex ante* determination of a strategy bundle, an *interim* strategy bundle (x^S, y^S) should be interpreted not as a contract among the players S, but as a set of contract offers to each member of S (the image of strategy (x^j, y^j), $\{(x^j(t^j), y^j(t^j)) \mid t^j \in T^j\}$, is the set of contracts offered to player j). They studied the *interim* Bayesian incentive-compatible core (definition 5.1.5) and *ex post* welfare loss.

There are three rounds of consecutive games in their framework: (1) an *interim* game played during the setup period, which determines a set of type-contingent contract offers; (2) choice of an *interim* contract from among the offered contracts during the setup period; and (3) an *interim* game played during the manufacturing period, which determines a re-contract. In all three games, Bayesian incentive compatibility plays a central role. The second game is actually not a genuine game but merely a collection of individual optimization problems, each division j choosing from the offered contracts one that is best for j. The third game is essentially a classical static game.

Ichiishi and Sertel's first result says that the assumptions in each of Ichiishi and Radner's existence theorems (theorems 9.2.5, 9.2.7 and 9.2.9) also guarantee the existence of a full-information revealing *interim* Bayesian incentive-compatible core plan for the first *interim* game. This is due to the following observation: By the weak Bayesian incentive compatibility which is implied by the Bayesian incentive compatibility, any strategy bundle $(x^N, y^N) \in \mathring{F}^{*N}$ satisfies

$$\forall\, t^j, t'^j \in T^j : E(x^j \mid t^j) = E(x^j \mid t'^j),$$

consequently,

$$\forall\, t^j \in T^j : E(x^j \mid t^j) = Ex^j.$$

Thus, the *interim* core and the *ex ante* core coincide in the profit-center game.

Their second result is the existence of a core re-contract of the third game, both for the convex case and for the case characterized by a structural relationship among the divisions (the [complementary supplier]-customer relationship). A re-contract improves welfare of each division. However, their example illustrates in the full-information-revealing case that although a re-contract removes the inefficiency that was caused by the Bayesian incentive compatibility, *ex post* efficiency can still not be achieved, due to the constraint still imposed by the information-pooling rule.

9.3 By Choosing a Contract

The next approach to information revelation borrows the idea from the principal-agent theory that agreeing or refusing to sign a contract reveals a private information. It applies to *interim* contracting in the private information case. There is no definitive written work based on this idea, however, and indeed it has a serious limitation if no other approaches are adopted concurrently. We will see two examples first in which a player's intention to sign a contract reveals his private information to the other players.

EXAMPLE 9.3.1 This example, attributed by Ichiishi and Sertel to an anonymous referee of their paper Ichiishi and Sertel (1998), describes a situation in which coalition formation is more difficult than is suggested by the coalitional stability condition of the *interim* Bayesian incentive-compatible strong equilibrium (definition 5.1.5). The essence of this example was observed by Wilson (1978, example 1, p. 809) when he illustrated the phenomenon of adverse selection which often violates opportunities for insurance.

Assume that each choice set is the real numbers, $C^j = \mathbf{R}$, and each utility function is the projection onto C^j, $u^j(c,t) = c^j$. Suppose the grand coalition is deliberating on the constant strategy bundle x:

$$\forall j \in N : \forall t \in T : x^j(t) = 1.$$

Suppose also that subcoalition $S := \{1,2\}$ finds the following strategy bundle $x'^S \in F^S(x)$: Assume $\pi(t) = \prod_{j \in N} \pi^j(t^j)$. Assume also for each $i \in S$, $T^i = \{H^i, L^i\}$, $\pi^i(H^i) = \pi^i(L^i) = 1/2$.

$$x'^1(t) := \begin{cases} 4, & \text{if } t = (H^1, L^2), \\ 0, & \text{otherwise;} \end{cases}$$

$$x'^2(t) := \begin{cases} 4, & \text{if } t = (L^1, H^2), \\ 0, & \text{otherwise;} \end{cases}$$

Notice that

$$E(x'^i \mid H^i) = 2 > 1 = E(x^i \mid H^i), \text{ for every } i \in S,$$

so S can improve upon x using x'^S when the true type profile is $\bar{t}^S = (H^1, H^2)$. However, player 1 knows that player 2 agrees to the joint strategy x'^S only when 2's true type is H^2, since

$$E(x'^2 \mid L^2) = 0 < 1 = E(x^2 \mid L^2).$$

Then player 2's agreement to x'^S reveals the information to player 1 that 2's true type is H^2. Given this information, player 1 does not agree to x'^S since

$$x'^1(t^1, H^2) = 0 < 1 = x^1(t^1, H^2), \text{ for } t^1 = H^1, L^1.$$

Thus, strategy x'^S cannot serve as a "blocking" strategy against x^S. □

EXAMPLE 9.3.2 This example, a variation of the previous example, describes a situation in which coalition formation is easier than is suggested by the coalitional stability condition of the *interim* Bayesian incentive-compatible strong equilibrium (definition 5.1.5). Assume again $C^j = \mathbf{R}$, $u^j(c,t) = c^j$, $S := \{1,2\}$, $T^i = \{H^i, L^i\}$, $\pi^i(H^i) = \pi^i(L^i) = 1/2$ for each $i \in S$. Suppose the grand coalition is deliberating on the constant strategy bundle x:

$$\forall\, j \in N : \forall\, t \in T : x^j(t) = 1.$$

Suppose also that coalition S finds the following strategy bundle $x''^S \in F^S(x)$:

$$x''^1(t) = x''^2(t) := \begin{cases} 1.5, & \text{if } t^2 = H^2, \\ 0, & \text{otherwise.} \end{cases}$$

Notice that

$$E(x''^1 \mid H^1) = E(x''^1 \mid L^1) = 0.75 < 1 = E(x^1 \mid H^1) = E(x^1 \mid L^1),$$

so S cannot improve upon x using x''^S according to the traditional "blocking" criterion. However, when the true type profile is $\bar{t}^S = (H^1, H^2)$, player 2 wants to agree to the joint strategy x''^S. When this happens, player 1 infers that 2's true type is H^2, so 1 also wants to agree to x''^S. Thus, strategy x''^S serves as a "blocking" strategy against x^S. □

The heart of these examples lie in the fact that players are comparing two strategy bundles. In example 9.3.2, if the members of coalition S decide to form their coalition and adopt strategy bundle x''^S when the grand coalition has been deliberating on strategy bundle x, it is because they received the information that event $\{(H^1, H^2), (L^1, H^2)\}$ has realized, and both players are better off with x''^S than with x given this additional information. It is important to keep in mind that this kind of information revelation occurs within a "blocking" coalition. The scenario here (in which two strategy bundles x''^S and x are compared) does not address how the private information is revealed only through the original strategy bundle x of the grand coalition N. In particular, given a strong equilibrium strategy bundle (or a core strategy bundle) x^*, this kind of information revelation does not occur, unless some coalition S explicitly tries to defect by adopting a strategy bundle x^S and experiences its inability to defect. We will see this last point in the next section.

Notice that strategies x'^S and x''^S in examples 9.3.1 and 9.3.2 are not $\mathcal{T}^j$-measurable or Bayesian incentive-compatible. Ichiishi and Sertel (1998), on the other hand, studied the profit-center game in which the measurability (postulate 9.1.2) and the Bayesian incentive compatibility (postulate 9.A.1) are strictly obeyed, and noted that information is not revealed through coalition formation, particularly due to the Bayesian incentive compatibility: For any strategy $(x^S, y^S) \in \hat{F}^{*S}$, $E(x^i \mid T^i)$ is constant on T for every $i \in S$. Then a "blocking" strategy would have to make every $i \in S$ better off (in terms of the *interim* expected imputation) for all possible type profiles. Thus, the fact that division i joins a particular coalition does not reveal any information to the other divisions of the coalition.

9.4 Update of the *Interim* Probabilities

Consider a Bayesian society,

$$\mathcal{S} := \left(\{C^j, T^j, u^j, \{\pi^j(\cdot \mid t^j)\}_{t^j \in T^j} \}_{j \in N}, \{\mathbf{C}_0^S, T(S), F^S\}_{S \in \mathcal{N}} \right)$$

(definition 2.1.3) in the private information case (section 3.2). Suppose that the grand coalition is entertaining a strategy bundle $\bar{x}$ which is private measurable (section 3.2) and Bayesian incentive-compatible (condition 4.1.1), but that the members of coalition S are contemplating in the *interim* period to defect and to take their own private measurable strategy bundle $x^S \in F'^S(\bar{x})$ after defection.

Holmström and Myerson (1983) formulated a mechanism to update the *interim* probabilities, $\pi^j(\cdot \mid t^j)$, $j \in S$, $t^j \in T^j$, through comparison of x^S and $\bar{x}$, as a specific instance of Selten's (1975) (trembling-hand) perfect equilibrium or of Kreps and Wilson's (1982) sequential equilibrium. We will adapt their basic idea to our present framework. We will obtain updated probabilities, regardless whether formation of the blocking coalition S turns out successful or unsuccessful.

Holmström and Myerson's scenario goes essentially as follows: There are two stages: In the first stage, the members of S announce simultaneously and independently whether to join the defecting coalition S by adopting strategy bundle x^S or to stay in the grand coalition N by keeping strategy bundle $\bar{x}$. If all members in S unanimously announce adoption of x^S, the defection is made definite, that is, coalition S is formed immediately after the first stage, and each member j takes action (choice) in accordance with x^j in the second stage; otherwise, coalition S fails to form, and each member j of the grand coalition N takes action in accordance with $\bar{x}^j$ in the second stage. There is a possibility of wrong action, that is, misrepresentation of a type, in the second stage. At the beginning of the second stage, the *interim* probabilities of the first stage, $\pi^j(\cdot \mid t^j)$, $j \in S$, $t^j \in T^j$, are updated. It is here that the concept of perfect equilibrium or of sequential equilibrium is applied, so Holmström and Myerson postulated that the players take behavior strategies (rather than pure strategies) throughout the scenario.

Notice that the term "pure strategy" in this scenario is a pair of (1) decision in the first stage as to joining S or staying in N, and (2) schedule of execution of x^S (or of $\bar{x}$) in the second stage contingent on the outcome of the first stage; it should not be confused with a pure strategy in the society S, such as x^j and $\bar{x}^j$. Due to players' behavior-strategy choice (as opposed to pure-strategy choice), the members S unanimously agree on adoption of x^S only probabilistically, given each type profile. Holmström and Myerson postulated that coalition S is indeed formed, if for each sequential equilibrium, there is a type profile at which, with a positive probability, the members S unanimously agree on adoption of x^S. The original strategy bundle $\bar{x}$, if feasible in the grand coalition, is called *durable* if no coalition S finds a strategy bundle x^S, with which S is formed. Compare this durability concept and the *interim* Bayesian incentive-compatible strong equilibrium concept (definition 5.1.5).

We begin formal presentation of the probability update. It is an extensive game, denoted by $\Gamma := \Gamma(\mathcal{S}, \bar{x}, x^S)$, in which the player set is given as N. We assume that each C^j (hence each X^j) is finite, and that $\pi^j(\cdot \mid t^j) \gg 0$ for all $j \in S$ and all $t^j \in T^j$ (hence $T(S) = T$).

At the origin of the extensive game Γ, the nature realizes a type profile $t \in T$. Then, the first stage of game Γ starts, in which the players j simultaneously and independently decide whether to adopt x^S and join the defecting coalition S or to keep $\bar{x}$ and stay in the grand coalition N. Player j, knowing his private information t^j but not knowing the others' types at this stage, has information set U_0^{j,t^j}; he has $\#T^j$ information sets at this stage, indexed by $t^j \in T^j$. At the information set U_0^{j,t^j}, player j decides whether to adopt x^S or to keep $\bar{x}$; these are his pure choices here. As a part of his behavior strategy, player j chooses a probability $r^j(t^j) \in [0,1]$ of adopting x^S at the information set U_0^{j,t^j} (so his probability of keeping $\bar{x}$ is $1 - r^j(t^j)$).

For the simplest nontrivial case in which

$$N = \{1,2\},$$
$$S = N,$$
$$T^j = \{t^j, t'^j\}, \ j \in N,$$

the game Γ is partially illustrated in figure 9.1. Here,

$$U_0^{1,t^1} = \{a,b\},$$
$$U_0^{1,t'^1} = \{c,d\},$$
$$U_0^{2,t^2} = \{e,f,i,j\},$$
$$U_0^{2,t'^2} = \{g,h,k,l\}.$$

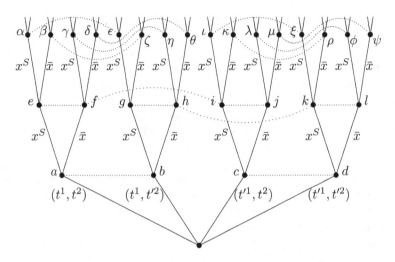

Figure 9.1: partial tree of the extensive game $\Gamma(\mathcal{S}, x^S, \bar{x})$

At information set U_0^{j,t^j}, j's possible pure choices are: to adopt x^S (indicated by the edge x^S), and to keep $\bar{x}$ (indicated by the edge $\bar{x}$).

Notice that U_0^{1,t^1} and U_0^{1,t'^1} are identified with $\{t^1\} \times T^2$ and $\{t'^1\} \times T^2$, respectively, and that U_0^{2,t^2} and U_0^{2,t'^2} are partitioned into the families,

$$U_0^{2,t^2}/\sim \;:=\; \{\{e,f\},\{i,j\}\},$$
$$U_0^{2,t'^2}/\sim \;:=\; \{\{g,h\},\{k,l\}\},$$

which are identified with $\{t^2\} \times T^1$ and $\{t'^2\} \times T^1$, respectively. Each partition is obtained by identifying player 1's choices that follow the same type profile.

In general, for every $j \in N$ and $t^j \in T^j$, information set U_0^{j,t^j} is partitioned into the family, $U_0^{j,t^j}/\sim$, by identifying the other players' choices that follow the same type profile, so this family is identified with $\{t^j\} \times T^{N\setminus\{j\}}$.

Suppose that a subset R of the players S chose to adopt x^S, and the rest $S \setminus R$ chose to keep $\bar{x}$ in the first stage. Then the second stage starts: Every player in S gets to know the colleagues' earlier actions, that is, each gets to know the fact that precisely R have chosen to adopt x^S. If all the players in S have unanimously chosen to adopt x^S (that is, if $R = S$), then the defecting coalition S is formed, and the members of S simultaneously execute x^S at this stage. If at least one player j has chosen to keep $\bar{x}$ (that is, if $S \setminus R \neq \emptyset$), then formation of S fails, and the players N simultaneously execute $\bar{x}$ at this stage. Each player $j \in S$, knowing his type t^j and the set R of the players who have expressed their intension to defect, is at the information set $U^{j,t^j}(R)$; he has $\#T^j \times 2^{\#S}$ information sets at this stage, indexed by $(t^j, R) \in T^j \times 2^S$. His pure choice at $U^{j,t^j}(R)$ is to present his type, which can be true or false. To present type $\tilde{t}^j$ means to take action $x^j(\tilde{t}^j)$ in case $R = S$, or to take action $\bar{x}^j(\tilde{t}^j)$ in case $R \neq S$. Define the space of all probabilities on T^j,

$$\Delta^{T^j} := \left\{ \sigma^j \in \mathbf{R}_+^{T^j} \,\Big|\, \sum_{t^j \in T^j} \sigma^j(t^j) = 1 \right\}.$$

As a part of his behavior strategy, player j chooses a probability $\sigma^j(\cdot \mid t^j, R) \in \Delta^{T^j}$ of presenting his type at the information set $U^{j,t^j}(R)$ (in case $R = S$, for example, he takes action $x^j(\tilde{t}^j)$ with probability $\sigma^j(\tilde{t}^j \mid t^j, S)$).

In the simple example of figure 9.1, player 1's information sets at the second stage are:

$$\begin{aligned}
U^{1,t^1}(\{1,2\}) &= \{\alpha, \epsilon\} \\
U^{1,t^1}(\{1\}) &= \{\beta, \zeta\} \\
U^{1,t^1}(\{2\}) &= \{\gamma, \eta\} \\
U^{1,t^1}(\emptyset) &= \{\delta, \theta\} \\
U^{1,t'^1}(\{1,2\}) &= \{\iota, \xi\} \\
U^{1,t'^1}(\{1\}) &= \{\kappa, \rho\} \\
U^{1,t'^1}(\{2\}) &= \{\lambda, \phi\} \\
U^{1,t'^1}(\emptyset) &= \{\mu, \psi\}
\end{aligned}$$

In game Γ, player j's behavior strategy is:

$$\beta^j := \left(\{r^j(t^j)\}_{t^j \in T^j}, \{\sigma^j(\cdot \mid t^j, R)\}_{(t^j, R) \in T^j \times 2^S} \right).$$

Suppose the players in S choose a behavior strategy bundle $\beta^S := \{\beta^j\}_{j \in S}$. Due to the Bayesian incentive compatibility of $\bar{x}$, the outsiders $N \setminus S$ execute $\bar{x}^{N \setminus S}$ truthfully when formation of S fails, since each of them $i \in N \setminus S$ is endowed with the *interim* probabilities $\pi^i(\cdot \mid t^i)$. Insider $j \in S$, given his private information t^j, has the *interim* probability $\pi^j(\cdot \mid t^j)$ as his probability estimate on the members of $U_0^{j,t^j}/\sim$. Denoting choice bundle $(x^i(t^i))_{i\in S}$ by $x^S(t^S)$, his expected utility given U_0^{j,t^j} is

$$Eu^j\left(\beta^S \mid U_0^{j,t^j}, \pi^j(\cdot \mid t^j)\right)$$

$$:= \sum_{s^{N\setminus\{j\}}\in T^{N\setminus\{j\}}} \pi^j(s^{N\setminus\{j\}} \mid t^j)$$

$$\times \left\{\prod_{i\in S\setminus\{j\}} r^i(s^i)\left[r^j(t^j)\sum_{s'^j}\sum_{s'^{S\setminus\{j\}}} \sigma^j(s'^j \mid t^j, S) \prod_{i\in S\setminus\{j\}} \sigma^i(s'^i \mid s^i, S)\right.\right.$$

$$\times u^j(x^j(s'^j), x^{S\setminus\{j\}}(s'^{S\setminus\{j\}}), \bar{x}^{N\setminus S}(s^{N\setminus S}), t^j, s^{N\setminus\{j\}})$$

$$+(1-r^j(t^j))\sum_{s'^j}\sum_{s'^{S\setminus\{j\}}} \sigma^j(s'^j \mid t^j, S\setminus\{j\}) \prod_{i\in S\setminus\{j\}} \sigma^i(s'^i \mid s^i, S\setminus\{j\})$$

$$\times u^j(\bar{x}^j(s'^j), \bar{x}^{S\setminus\{j\}}(s'^{S\setminus\{j\}}), \bar{x}^{N\setminus S}(s^{N\setminus S}), t^j, s^{N\setminus\{j\}})]$$

$$+ \sum_{R\subsetneq S\setminus\{j\}} \left[\left(\prod_{i\in R\cup\{j\}} r^i(s^i)\right)\left(\prod_{i\in S\setminus(R\cup\{j\})} (1-r^i(s^i))\right)\right.$$

$$\times \sum_{s'^j}\sum_{s'^{S\setminus\{j\}}} \sigma^j(s'^j \mid t^j, R\cup\{j\}) \prod_{i\in S\setminus\{j\}} \sigma^i(s'^i \mid s^i, R\cup\{j\})$$

$$\times u^j(\bar{x}^j(s'^j), \bar{x}^{S\setminus\{j\}}(s'^{S\setminus\{j\}}), \bar{x}^{N\setminus S}(s^{N\setminus S}), t^j, s^{N\setminus\{j\}})]$$

$$+ \left[\left(\prod_{i\in R} r^i(s^i)\right)\left(\prod_{i\in S\setminus R}(1-r^i(s^i))\right)\right.$$

$$\times \sum_{s'^j}\sum_{s'^{S\setminus\{j\}}} \sigma^j(s'^j \mid t^j, R) \prod_{i\in S\setminus\{j\}} \sigma^i(s'^i \mid s^i, R)$$

$$\times u^j(\bar{x}^j(s'^j), \bar{x}^{S\setminus\{j\}}(s'^{S\setminus\{j\}}), \bar{x}^{N\setminus S}(s^{N\setminus S}), t^j, s^{N\setminus\{j\}})]\}.$$

Suppose information set $U^{j,t^j}(R)$ is reached. The set $U^{j,t^j}(R)$ can be partitioned into a family of classes, $U^{j,t^j}(R)/\sim$, by identifying the other players' choices in the second stage that follow the same type profile and R. This family is identified with $\{t^j\} \times T^{N\setminus\{j\}}$. Denote by $q^j(\cdot \mid t^j, R)$ player j's updated probability on the members of $U^{j,t^j}(R)/\sim$; we will present Selten's (1975) formula for this probability in definition 9.4.1 and the paragraph preceding it. Then, his expected utility given $U^{j,t^j}(R)$ is: If $R = S$,

$$Eu^j\left(\beta^S \mid U^{j,t^j}(S), q^j(\cdot \mid t^j, S)\right)$$

$$:= \sum_{s^{N\setminus\{j\}} \in T^{N\setminus\{j\}}} q^j(s^{N\setminus\{j\}} \mid t^j, S)$$

$$\times \sum_{s'^j} \sum_{s'^{S\setminus\{j\}}} \sigma^j(s'^j \mid t^j, S) \prod_{i \in S\setminus\{j\}} \sigma^i(s'^i \mid s^i, S)$$

$$\times u^j(x^j(s'^j), x^{S\setminus\{j\}}(s'^{S\setminus\{j\}}), \bar{x}^{N\setminus S}(s^{N\setminus S}), t^j, s^{N\setminus\{j\}});$$

If $R \neq S$,

$$Eu^j\left(\beta^S \;\middle|\; U^{j,t^j}(R), q^j(\cdot \mid t^j, R)\right)$$

$$:= \sum_{s^{N\setminus\{j\}} \in T^{N\setminus\{j\}}} q^j(s^{N\setminus\{j\}} \mid t^j, R)$$

$$\times \sum_{s'^j} \sum_{s'^{S\setminus\{j\}}} \sigma^j(s'^j \mid t^j, R) \prod_{i \in S\setminus\{j\}} \sigma^i(s'^i \mid s^i, R)$$

$$\times u^j(\bar{x}^j(s'^j), \bar{x}^{S\setminus\{j\}}(s'^{S\setminus\{j\}}), \bar{x}^{N\setminus S}(s^{N\setminus S}), t^j, s^{N\setminus\{j\}}).$$

Here, the outsiders $i \in N \setminus S$, having been left out from the first step of game $\Gamma(S, \bar{x}, x^S)$, cannot update his probability, so maintain their *interim* probabilities $\pi^i(\cdot \mid s^i)$ when formation of S fails. Since $\bar{x}$ is Bayesian incentive-compatible, they execute $\bar{x}^i$ truthfully.

We remark that given $\left(U^{j,t^j}(R), q^j(\cdot \mid t^j, R)\right)$, j's conditional expected utility $Eu^j\left(\beta^S \;\middle|\; U^{j,t^j}(R), q^j(\cdot \mid t^j, R)\right)$ depends only on

$$\left(\sigma^j(\cdot \mid t^j, R), \{\sigma^i(\cdot \mid s^i, R)\}_{i \in S\setminus\{j\}, s^{S\setminus\{j\}} \in T^{S\setminus\{j\}}}\right),$$

rather than fully on β^S.

If $\beta \gg \mathbf{0}$ and $1 - r^j(t^j) > 0$ for all $j \in N$ and all $t^j \in T^j$, then player j's updated probability $q^j(\cdot \mid t^j, R)$ on the members of $U^{j,t^j}(R)/\sim$ is computed as follows: Define

$$p(r^{S\setminus\{j\}}(s^{S\setminus\{j\}}), R) := \left(\prod_{i \in R\setminus\{j\}} r^i(s^i)\right)\left(\prod_{i \in S\setminus(R\cup\{j\})} (1 - r^i(s^i))\right).$$

Then the required probability is given by

$$q^j(s^{N\setminus\{j\}} \mid t^j, R)$$
$$= \frac{\pi^j(s^{N\setminus\{j\}} \mid t^j) p(r^{S\setminus\{j\}}(s^{S\setminus\{j\}}), R)}{\sum_{s'^{N\setminus\{j\}} \in T'^{N\setminus\{j\}}} \pi^j(s'^{N\setminus\{j\}} \mid t^j) p(r^{S\setminus\{j\}}(s'^{S\setminus\{j\}}), R)}.$$

However, some $r^i(s'^i)$ may be equal to 0 or to 1, in which case this formula for $q^j(\cdot \mid t^j, R)$ may not be applicable. To overcome this difficulty, Selten considered a sequence of strictly positive behavior strategy bundles which converges to β. For each $\nu \in \mathbf{N}$, define a trimmed set of j's behavior strategies,

$$B^j_\nu := \left\{ \beta^j \;\middle|\; \begin{array}{l} \forall\, t^j \in T^j : \frac{1}{\nu} \leq r^j(t^j) \leq 1 - \frac{1}{\nu}, \\ \forall\, R \in 2^S : \forall\, s^j \in T^j : \frac{1}{\nu} \leq \sigma^j(s^j \mid t^j, R) \end{array} \right\}.$$

An *agent* in game $\Gamma(\mathcal{S}, \bar{x}, x^S)$ is a pair (j, t^j) of player j and his type t^j. Condition (iii) in the following definition 9.4.1 means that when the agents $\{(j, t^j) \mid j \in S, t^j \in T^j\}$ play the game $\Gamma(\mathcal{S}, \bar{x}, x^S)$, the behavior strategy bundle $\{r^{*j}, \{\sigma^{*j}(\cdot \mid t^j, R)\}_{R \in 2^R}\}_{(j,t^j)}$ constitutes a Nash equilibrium. Condition (iv) is a perfectness criterion: Agent (j, t^j) at each information set $U^{j,t^j}(R)$ responds to the others' choices $\sigma^{*i}(\cdot \mid s^i, R), i \in S\setminus\{j\}, s^i \in T^i$, optimally relative to the updated probability $q^{*j}(\cdot \mid t^j, R)$. Notice that (i), (ii), and (v) together imply (iii) and (iv).

DEFINITION 9.4.1 (Selten, 1975; Kreps and Wilson, 1982) A behavior strategy bundle $\{\beta^{*j}\}_{j \in S}$ of game $\Gamma(\mathcal{S}, \bar{x}, x^S)$ and probabilities $q^{*j}(\cdot \mid t^j, R)$ on $U^{j,t^j}(R)/\sim$, $j \in S, t^j \in T^j, R \in 2^S$, are called a *sequential equilibrium* of $\Gamma(\mathcal{S}, \bar{x}, x^S)$, if there exists a sequence of behavior strategy bundles $\{\{\beta^{\nu j}\}_{j \in S}\}_\nu$, $\beta^{\nu j} \in B^j_\nu$, such that for all $j \in S$,
(i) $\beta^{\nu j} \to \beta^{*j}$, as $\nu \to \infty$;
(ii) $q^{*j}(\cdot \mid t^j, R) = \lim_{\nu \to \infty} q^{\nu j}(\cdot \mid t^j, R)$, where

$$q^{\nu j}(s^{N\setminus\{j\}} \mid t^j, R)$$
$$= \frac{\pi^j(s^{N\setminus\{j\}} \mid t^j) p(r^{\nu S\setminus\{j\}}(s^{S\setminus\{j\}}), R)}{\sum_{s'^{N\setminus\{j\}} \in T'^{N\setminus\{j\}}} \pi^j(s'^{N\setminus\{j\}} \mid t^j) p(r^{\nu S\setminus\{j\}}(s'^{S\setminus\{j\}}), R)};$$

(iii) $\forall\, t^j \in T^j, \forall\, \beta^j$,

$$Eu^j\left(\beta^{*S} \;\middle|\; U^{j,t^j}_0, \pi^j(\cdot \mid t^j)\right) \geq Eu^j\left(\beta^j, \beta^{*S\setminus\{j\}} \;\middle|\; U^{j,t^j}_0, \pi^j(\cdot \mid t^j)\right);$$

(iv) $\forall\, t^j \in T^j, \forall\, R \in 2^S, \forall\, \sigma^j(\cdot \mid t^j, R)$,

$$Eu^j\left(\sigma^{*j}(\cdot \mid t^j, R), \{\sigma^{*i}(\cdot \mid s^i, R)\}_{i, s^{S\setminus\{j\}}} \;\middle|\; U^{j,t^j}(R), q^j(\cdot \mid t^j, R)\right)$$
$$\geq Eu^j\left(\sigma^j(\cdot \mid t^j, R), \{\sigma^{*i}(\cdot \mid s^i, R)\}_{i, s^{S\setminus\{j\}}} \;\middle|\; U^{j,t^j}(R), q^j(\cdot \mid t^j, R)\right).$$

The pair $(\{\beta^{*j}\}_{j \in S}, \{q^{*j}(\cdot \mid t^j, R)\}_{j \in S, t^j \in T^j, R \in 2^S})$ is called a *perfect equilibrium* of $\Gamma(\mathcal{S}, \bar{x}, x^S)$, if (i) and (ii) are satisfied and if
(v) the sequence $\{\{\beta^{\nu j}\}_{j \in S}\}_\nu$ can be taken so that for each ν, $\{\beta^{\nu j}\}_{j \in S}$ is a Nash equilibrium of the game in which each player j's behavior strategies are restricted to B^j_ν.

A perfect equilibrium of $\Gamma(\mathcal{S}, \bar{x}, x^S)$ always exists, hence so does a sequential equilibrium. We quote Selten's existence result:

THEOREM 9.4.2 (Selten, 1975) *Let $\mathcal{S}$ be a Bayesian society, such that C^j is finite, and $\pi^j(\cdot \mid t^j) \gg 0$ for all $j \in N$ and all $t^j \in T^j$. Let $\bar{x} : T \to C$ and $x^S : T \to C^S$ be private-measurable strategy bundles. Then there exists a perfect equilibrium of $\Gamma(\mathcal{S}, \bar{x}, x^S)$ in behavior strategies.*

REMARK 9.4.3 In defining a sequential or perfect equilibrium of game $\Gamma(\mathcal{S}, \bar{x}, x^S)$, a sequence of behavior strategy bundles $\{\{\beta^{\nu j}\}_{j \in S}\}_\nu$, $\beta^{\nu j} \in B^j_\nu$, satisfying condition (i) of definition 9.4.1 is used, only in order to define the updated probabilities $q^{*j}(\cdot \mid t^j, R)$, $j \in S$, $t^j \in T^j$, $R \in 2^S$. For this purpose, the trimmed set B^j_ν can be replaced by a simpler set,

$$\left\{ \beta^j \;\middle|\; \forall\, t^j \in T^j : \frac{1}{\nu} \leq r^j(t^j) \leq 1 - \frac{1}{\nu} \right\}.$$

□

Let $(\{\beta^{*j}\}_{j \in S}, \{q^{*j}(\cdot \mid t^j, R)\}_{j \in S, t^j \in T^j, R \in 2^S})$ be a sequential equilibrium or a perfect equilibrium of $\Gamma(\mathcal{S}, \bar{x}, x^S)$. Player j of type t^j knows that at this equilibrium, information set $U^{j,t^j}(R)$ is reached with probability

$$\sum_{s^{N \setminus \{j\}} \in T'^{N \setminus \{j\}}} \pi^j(s^{N \setminus \{j\}} \mid t^j) \left(\prod_{i \in R} r^{*i}(s^i) \right) \left(\prod_{i \in S \setminus R} (1 - r^{*i}(s^i)) \right).$$

He knows, therefore, that with this probability, his *interim* probability $\pi^j(\cdot \mid t^j)$ on $T^{N \setminus \{j\}}$ is updated to probability $q^{*j}(\cdot \mid t^j, R)$.

The preceding paragraph presented the explanatory power of the sequential or perfect equilibrium concept for game $\Gamma(\mathcal{S}, \bar{x}, x^S)$. However, the concept is too weak to serve as a *feasible* solution to the question of whether or not the defecting coalition S will be actually formed and realize x^S. To see this point, let $x^S \in F'^S(\bar{x})$, and consider a pure strategy bundle in $\Gamma(\mathcal{S}, \bar{x}, x^S)$, in which precisely all the players S announce adoption of x^S, and each player $j \in S$ reports type $\tilde{t}^j(t^j, S) \in T^j$ when his true type is t^j. For this pure strategy bundle of $\Gamma(\mathcal{S}, \bar{x}, x^S)$ to be *feasible* in the context of $\mathcal{S}$, we require that the strategy bundle in $\mathcal{S}$ defined by

$$t \mapsto \left\{ x^i \left(\tilde{t}^i(t^i, S) \right) \right\}_{i \in S},$$

is a member of $F'^S(\bar{x})$. (Recall the argument in the last seven paragraphs of section 4.1, which makes the same point.) The feasibility question becomes obviously more involved for a behavior strategy bundle. The feasibility question occurs because of the rich structure of a Bayesian society, in which F^S is arbitrarily given *a priori*. It does not occur in Holmström and Myerson' model, which essentially postulates that $F^S(\bar{x}) = \left(C_0^S(\bar{x})\right)^T$ for some $C_0^S(\bar{x}) \subset C^S$. We will see in the paragraph following condition 9.4.4 below how this question is resolved.

We have emphasized in section 4.1 that the members of a coalition agree on a strategy bundle in order to plan a best choice bundle preparing for every contingency, so it is essential that choices are later made exactly as scheduled. This is the reason why we postulate throughout this book that each coalition designs only Bayesian incentive-compatible strategy bundles. We are postulating the same here. In the basic model of section 4.1, however, Bayesian incentive compatibility was defined relative to the *interim* probabilities, $\pi^j(\cdot \mid t^j)$, $j \in S$, $t^j \in T^j$, since for every (j, t^j), probability $\pi^j(\cdot \mid t^j)$ was used in computing the conditional expected utility of a strategy bundle given t^j. According to the present scenario of game $\Gamma(\mathcal{S}, \bar{x}, x^S)$ on the other hand, player j reports his type at his information set $U^{j,t^j}(R)$ when he has the updated probability $q^j(\cdot \mid t^j, R)$ (rather than $\pi^j(\cdot \mid t^j)$). Therefore, the Bayesian incentive compatibility condition needs to be modified in the present context:

CONDITION 9.4.4 Let $\mathcal{S}$ be a Bayesian society, such that C^j is finite, and $\pi^j(\cdot \mid t^j) \gg \mathbf{0}$ for all $j \in N$ and all $t^j \in T^j$, and consider the private information case. Choose any private measurable, Bayesian incentive-compatible strategy bundle $\bar{x} : T \to C$, and any private measurable strategy bundle $x^S \in F'^S(\bar{x})$. The two bundles $(\bar{x}, x^S)$ are called a *Bayesian incentive-compatible strategy bundle pair*, if for any sequential equilibrium $\left(\{\beta^{*j}\}_{j \in S}, \{q^{*j}(\cdot \mid t^j, R)\}_{j \in S, t^j \in T^j, R \in 2^S}\right)$ of $\Gamma(\mathcal{S}, \bar{x}, x^S)$, it follows that

$$\forall j \in S : \forall t^j \in T^j : \forall R \in 2^S :$$

$$\sigma^{*j}(s^j \mid t^j, R) = \begin{cases} 1 & \text{if} & s^j = t^j \\ 0 & \text{for all} & s^j \in T^j \setminus \{t^j\}. \end{cases}$$

For a Bayesian incentive-compatible strategy bundle pair $(\bar{x}, x^S)$, a sequential or perfect equilibrium of game $\Gamma(\mathcal{S}, \bar{x}, x^S)$ is feasible in the context of $\mathcal{S}$ (so that the feasibility question raised in the second paragraph preceding condition 9.4.4 is positively resolved), since $x^j(t^j)$ ($\bar{x}^j(t^j)$, resp.) is truthfully chosen at information set $U^{j,t^j}(S)$ (at $U^{j,t^j}(R)$, $R \neq S$, resp.).

For each private measurable strategy bundle $\bar{x} : T \to C$ and each $S \in \mathcal{N}$, define

$$\hat{F}^{\dagger S}(\bar{x}) := \left\{ x^S \in F'^S(\bar{x}) \;\middle|\; \begin{array}{l} (\bar{x}, x^S) \text{ is a Bayesian incentive-} \\ \text{compatible strategy bundle pair.} \end{array} \right\}.$$

9: Information Revelation

The following is a descriptive solution concept for $\mathcal{S}$; it extends the *interim* Bayesian incentive-compatible strong equilibrium concept (definition 5.1.5) to the situation in which *interim* probabilities $\pi^j(\cdot \mid t^j)$ are updated during the deliberation period for any defecting coalition S, when each member of S signs up either the grand coalition's strategy bundle x^* or S's alternative strategy bundle x^S.

DEFINITION 9.4.5 Let $\mathcal{S}$ be a Bayesian society, such that C^j is finite, and $\pi^j(\cdot \mid t^j) \gg 0$ for all $j \in N$ and all $t^j \in T^j$, and consider the private information case. A strategy bundle $x^* : T \to C$ is called *durable*, if
(i) $x^* \in \hat{F}^N(x^*)$; and
(ii) it is not true that

$$\exists\, S \in \mathcal{N} : \exists\, x^S \in \hat{F}^{\dagger S}(x^*) :$$
$$(\forall\, \beta^{*S} : \text{ sequential equilibrium of } \Gamma(\mathcal{S}, x^*, x^S)) :$$
$$\exists\, t \in T : \prod_{i \in S} r^{*i}(t^i) > 0.$$

The general existence theorem for a durable strategy bundle is still an open question. A more fundamental open question is to build a theory of *interim*-probability update without resorting to behavior strategies in game $\Gamma(\mathcal{S}, \bar{x}, x^S)$: We would like to explain exact action of each player. Indeed, while actions are observable (of course, we are excluding moral hazard), probabilistic choices are impossible to observe, which raises the usual serious philosophical question on the validity of the postulate that each player j takes the others' *unobservable* behavior strategies $\beta^{S \setminus \{j\}}$ as given.

REMARK 9.4.6 There is a difference between Holmström and Myerson's treatment of durability and the present section's treatment of durability. In Holmström and Myerson's treatment, there is no update of the *interim* probability when any of the information sets $U^{j,t^j}(R)$, $R \neq S$, is reached (they consider the probability update only given the information set $U^{j,t^j}(S)$). So the players in S do not learn anything out of their failure to form a defecting coalition. In the present treatment on the other hand, players learn from comparison of $\bar{x}$ and x^S, regardless whether S is indeed formed or not. Since the conditional expected utility given $U^{j,t^j}(R)$ for some $R \neq S$ is computed with respect to the updated probability, we need to re-formulate the Bayesian incentive compatibility accordingly. □

9.5 By Credible Transmission of Information During the Contract Negotiation

This approach, taken by Yazar (2001), endogenously determines a communication system as a part of coalition's strategy bundle during the *ex ante* period of strategy negotiation. Recall her formulation of a strategy in the Bayesian pure exchange economy (section 4.3) and her Bayesian incentive compatibility condition 4.3.1. For each coalition S, denote by $\hat{F}^S$ the set of all measurable, Bayesian incentive-compatible and attainable strategies,

$$\left\{ \{z^j, \mathcal{C}^j\}_{j \in S} \ \middle| \ \begin{array}{l} \forall \, j \in S : z^j \text{ is } \bigvee_{i \in S} \mathcal{C}^i\text{-measurable, and } \mathcal{C}^j \subset \mathcal{T}^j, \\ \{z^j, \mathcal{C}^j\}_{j \in S} \text{ is Bayesian incentive-compatible,} \\ \forall \, j \in S : z^j \geq -e^j, \\ \forall \, t \in T : \sum_{j \in S} z^j(t) = \mathbf{0} \end{array} \right\}.$$

Here, as in section 4.3, we keep assuming $T(\pi^j) = T(S)$ for every $j \in S$, for simplicity of analysis. A strategy bundle $\{z^{*j}, \mathcal{C}^{*j}\}_{j \in N}$ of the grand coalition in Bayesian pure exchange economy $\mathcal{E}_{pe}$ is said to be in the *EC-core (endogenous communication plan core)*, (i) if it is in $\hat{F}^N$, and (ii) if it is not true that there exist $S \in \mathcal{N}$ and $\{z^j, \mathcal{C}^j\}_{j \in S} \in \hat{F}^S$ such that $Eu^j(z^j + e^j) > Eu^j(z^{*j} + e^j)$ for every $j \in S$. The communication system $\{\mathcal{A}^{*j}\}_{j \in N}$, $\mathcal{A}^{*j} := \mathcal{T}^j \bigvee (\bigvee_{i \in N} \mathcal{C}^{*i})$, sustains as a result of credible talk at the contract negotiation.

Yazar's main result (theorem 9.5.2) follows immediately from her lemma on nested structures:

LEMMA 9.5.1 (Yazar, 2001) *For any coalition $S \in \mathcal{N}$, let $\{\mathcal{C}^j\}_{j \in S}$ and $\{\mathcal{C}'^j\}_{j \in S}$ be two communication plan bundles, and let $\{z^j\}_{j \in S}$ be a net trade bundle. If $\mathcal{C}'^j \subset \mathcal{C}^j$ for every $j \in S$ and if $\{z^j, \mathcal{C}'^j\}_{j \in S} \in \hat{F}^S$, then $\{z^j, \mathcal{C}^j\}_{j \in S} \in \hat{F}^S$.*

THEOREM 9.5.2 (Yazar, 2001) *Let $\{\mathcal{C}^j\}_{j \in N}$ and $\{\mathcal{C}'^j\}_{j \in N}$ be two communication plan bundles, and let $\{z^j\}_{j \in N}$ be a net trade bundle for the grand coalition. If $\mathcal{C}'^j \subset \mathcal{C}^j$ for every $j \in N$ and if $\{z^j, \mathcal{C}'^j\}_{j \in N}$ is in the EC-core, then $\{z^j, \mathcal{C}^j\}_{j \in N}$ is also in the EC-core.*

In particular, if the EC-core is nonempty at all, then there exists a strategy bundle in the EC-core which gives rise to the full communication system. This result agrees with the existence of a full-information revealing Bayesian incentive-compatible strong equilibrium, generic or exact, according to the approach to information revelation via contract execution (theorems 9.1.4, 9.1.6, 9.2.5, 9.2.7, 9.2.9).

9: Information Revelation

Proof of Lemma 9.5.1 Let $C'^j \subset C^j$ for every $j \in S$. Denote by $\mathcal{P}'^j$ ($\mathcal{P}^j$, resp.) the partition of T consisting of the minimal elements of C'^j (C^j, resp.). Suppose that $\{z^j, C'^j\}_{j\in S} \in \hat{F}^S$, but that $\{z^j, C^j\}_{j\in S} \notin \hat{F}^S$. Then,

$$\exists j \in S : \exists \bar{t} \in T(S) : \exists \bar{C}^j \in \mathcal{P}^j : \forall t \in \{\bar{t}^j\} \times T^{N\setminus\{j\}} \bigcap T(S) :$$
$$Eu^j(z^j(\bar{E} \bigcap T(S)) + e^j(\bar{t}^j) \mid \bar{t}^j) > Eu^j(z^j(t) + e^j(\bar{t}^j) \mid \bar{t}^j),$$

where $\bar{E} := \bar{C}^j \times \prod_{i\in S\setminus\{j\}} C^i(\bar{t}^i)$. Notice that there is $\bar{E}' := \bar{C}'^j \times \prod_{i\in S\setminus\{j\}} C'^i(\bar{t}^i) \in \bigcap_{i\in S}\mathcal{P}'^i$ such that $\bar{E}' \supset \bar{E}$.

It suffices to show at $\bar{t}^j$ that (1) $\bar{C}'^j$ is false information, and (2) by passing $\bar{C}'^j$ on to the others, that is, by acting $z^j(\bar{C}'^j)$ rather than acting $z^j(\bar{t})$, consumer j gets a higher *interim* expected utility given $\bar{t}^j$, since this would contradict the hypothesis $\{z^j, C'^j\}_{j\in S} \in \hat{F}^S$.

But z^j is $\bigvee_{i\in S} C'^i$-measurable, so it is constant not only on $\bar{E}$ but also on $\bar{E}'$. Therefore,

$$Eu^j(z^j(\bar{E}' \bigcap T(S)) + e^j(\bar{t}^j) \mid \bar{t}^j)$$
$$= Eu^j(z^j(\bar{E} \bigcap T(S)) + e^j(\bar{t}^j) \mid \bar{t}^j)$$
$$> Eu^j(z^j(t) + e^j(\bar{t}^j) \mid \bar{t}^j).$$

□

For the special case in which each utility function $u^j(\cdot, t)$ is affine linear on the consumption set $\mathbf{R}^l_+$, Yazar (2001) also established nonemptiness of the EC-core by direct application of Scarf's theorem for nonemptiness of the core (theorem 8.A.2 of this book) to the non-side-payment game defined by

$$V(S) := \left\{ u \in \mathbf{R}^N \;\middle|\; \begin{array}{l} \exists \{z^j, C^j\}_{j\in S} \in \hat{F}^S : \\ \forall j \in S : u_j \leq Eu^j(z^j + e^j) \end{array} \right\}.$$

For Vohra's Bayesian incentive compatibility (condition 4.2.1) applied to an arbitrary communication plan, the analogue of lemma 9.5.1 is trivially true. We may, therefore, assume *without loss of generality* that coalition S designs a strategy bundle $\{z^j, \mathcal{T}^j\}_{j\in S}$ with the full communication plan.

9.A Appendix to Chapter 9

The purpose of this appendix is three-fold. The first purpose is to present a precise definition of Bayesian incentive compatibility for the two-*interim*-period model of Bayesian society which satisfies the information-revelation process (section 9.1), and to establish its characterization. The second purpose is to prove theorem 9.1.6. The third purpose is to prove theorem 9.2.7.

The required Bayesian incentive compatibility condition extends d'Aspremont and Gérard-Varet's (1979) original definition (condition 4.1.1) to the present context. It also incorporates Murat Sertel's idea, *pretend-but-perform principle*, later developed by Koray and Sertel (see, e.g., Koray and Sertel, 1992). According to this principle, players are allowed to *pretend* to have their chosen types, *but* must thereafter *perform* so as not to belie them.

A *pretension function* of player j is a function $\sigma : T^j \to T^j$, which says that when his true type is t^j, he acts (makes a choice) as though his type were $\sigma(t^j)$. Given any algebra $\mathcal{B}^j$ on T^j, denote by $\text{endo}(T^j, \mathcal{B}^j)$ the set[5] of all functions that map each t^j into the minimal set of $\mathcal{B}^j$ that contains t^j. If every member of coalition S has information structure $\mathcal{B}^j$ about player j's type, then j can only choose a pretension function $\sigma \in \text{endo}(T^j, \mathcal{B}^j)$.

In order to define the Bayesian incentive compatibility, one needs to clarify first which choice of player j is *legal* in the sense that the other members of the coalition S cannot catch j's false pretension about his true type. Suppose the members of S are deliberating on whether or not to sign a contract $x^S \in F'^S(\bar{x})$. At the beginning of the first *interim* period, player j's information structure is given as T^j, and no other member has any part of this information (that is, if $i \in S \setminus \{j\}$, then $T^i \cap T^j = \{\emptyset, T\}$). So, player j is not caught in the first month no matter which choice he makes from $\{x_1^j(t^j) \mid t^j \in T^j\}$; that is, he can make choice according to any pretension function $\sigma \in \text{endo}(T^j, \{\emptyset, T^j\})$, so that when player j's true type is t^j, he makes the choice $x_1^j(\sigma(t^j))$. By acting according to the function $x_1^j \circ \sigma$, player j having his true type t^j passes on to all the other members of S the information that event $A := (x_1^j)^{-1}(x_1^j \circ \sigma(t^j))$ has occurred. This information may be false, that is, t^j may not be a member of A, but the other members take it as j's testimony about himself and expect that j will act according to this information in the second *interim* period, that is, j will have to make a choice from $x_2^j(A)$ in the second month. Therefore, j's pretension function in the second month has to be of the form $\tau \circ \sigma$ for some $\tau \in \text{endo}(T^j, \mathcal{A}(x_1^j))$.

When j chooses such a pair of pretension functions, $\sigma \in \text{endo}(T^j, \{\emptyset, T^j\})$ and $\tau \in \text{endo}(T^j, \mathcal{A}(x_1^j))$, the other members $i \in S \setminus \{j\}$, acting honestly, would make choices $(x_1^i(t^i), x_2^i(\sigma(t^j), t^{S \setminus \{j\}}))$, because x_2^i is $\hat{\mathcal{T}}^i(x_1^S)$-

---
[5]Let $\mathcal{P}$ be the partition of T^j that consists of the minimal nonempty members of $\mathcal{B}^j$. Then, $\text{endo}(T^j, \mathcal{B}^j) := \{\sigma : T^j \to T^j \mid \forall P \in \mathcal{P} : \sigma(P) \subset P\}$.

9: Information Revelation

measurable and

$$(\sigma(t^j), t^{S\setminus\{j\}}) \in ((x_1^j)^{-1}(x_1^j \circ \sigma(t^j)), (x_1^{S\setminus\{j\}})^{-1}(x_1^{S\setminus\{j\}}(t^{S\setminus\{j\}}))).$$

The present concept of *Bayesian incentive compatibility* says that player j cannot benefit from any pair of pretension functions that are not caught. Recall that $F'^S(\bar{x})$ is the set of all $x^S \in F^S(\bar{x})$ such that each x_1^j is T^j-measurable and x_2^j is $\hat{T}^j(x_1^S)$-measurable.

POSTULATE 9.A.1 Let S be a two-*interim*-period Bayesian society (society which satisfies postulates 9.1.1 and 9.1.2). Suppose that the grand coalition is entertaining a strategy bundle $\bar{x}$, but that the members of coalition S are contemplating to defect and to take their own strategy bundle after defection. The members of S agree only on those strategies $x^S \in F'^S(\bar{x})$ that are *Bayesian incentive-compatible*, that is,

$$\forall\, j \in S; \forall\, \sigma \in \mathrm{endo}(T^j, \{\emptyset, T^j\}) : \forall\, \tau \in \mathrm{endo}(T^j, \mathcal{A}(x_1^j)) : \forall\, t \in T :$$
$$Eu^j(x^S, \bar{x}^{N\setminus S} \mid \hat{T}^j(x_1^S))(t)$$
$$\geq\; Eu^j\left((x_1^j \circ \sigma,\, x_2^j \circ (\tau \circ \sigma,\, \mathrm{id})),\, (x_1^{S\setminus\{j\}},\, x_2^{S\setminus\{j\}} \circ (\sigma,\, \mathrm{id})),\right.$$
$$\left.\bar{x}^{N\setminus S} \;\middle|\; \hat{T}^j(x_1^S)\right)(t),$$

where id is the identity map on $T^{S\setminus\{j\}}$.

We can simplify this condition: Observe that for any $x^S \in F'^S(\bar{x})$, $\sigma \in \mathrm{endo}(T^j, \{\emptyset, T^j\})$, and $\tau \in \mathrm{endo}(T^j, \mathcal{A}(x_1^j))$, it follows that

$$\forall\, t^S \in T^S : x_1^j \circ \sigma(t^j) = x_1^j \circ \tau \circ \sigma(t^j),$$

and

$$\forall\, i \in S \setminus \{j\} : x_2^i(\sigma(t^j), t^{S\setminus\{j\}}) = x_2^i(\tau \circ \sigma(t^j), t^{S\setminus\{j\}}).$$

The first identity is due to $\tau \in \mathrm{endo}(T^j, \mathcal{A}(x_1^j))$. The second identity is due to $\tau \in \mathrm{endo}(T^j, \mathcal{A}(x_1^j))$ and the fact that x_2^i is $T^i(x_1^S)$-measurable. Consequently, we establish the following simplification:

FACT 9.A.2 *Strategy bundle x^S is in $\hat{F}^S(\bar{x})$, iff $x^S \in F'^S(\bar{x})$, and*

$$\forall\, j \in S : \forall\, \sigma \in \mathrm{endo}(T^j, \{\emptyset, T^j\}) : \forall\, t \in T :$$
$$Eu^j(x^S, \bar{x}^{N\setminus S} \mid \hat{T}^j(x_1^S))(t)$$
$$\geq\; Eu^j\left((x_1^j \circ \sigma,\, x_2^j \circ (\sigma,\, \mathrm{id})),\, (x_1^{S\setminus\{j\}},\, x_2^{S\setminus\{j\}} \circ (\sigma,\, \mathrm{id})),\right.$$
$$\left.\bar{x}^{N\setminus S} \;\middle|\; \hat{T}^j(x_1^S)\right)(t),$$

where id is the identity map on $T^{S\setminus\{j\}}$.

In the light of this simplification, the following facts 9.A.3 and 9.A.4 are immediate. Fact 9.A.5 is also straightforward.

FACT 9.A.3 *For each $\bar{x} \in X$, $\hat{F}^S(\bar{x}) \subset \hat{G}_w^S(\bar{x})$, and $\hat{G}_s^S(\bar{x}) \subset \hat{G}_w^S(\bar{x})$.*

FACT 9.A.4 *If $x^S \in G^S(\bar{x})$ and if x_1^j is 1-1 on T^j for every $j \in S$, then $x^S \in \hat{F}^S(\bar{x})$ iff $x^S \in \hat{G}_s^S(\bar{x})$.*

FACT 9.A.5 *Under assumptions 1 and 3, for each $S \in \mathcal{N}$, sets $\hat{G}_s^S(\bar{x})$ and $\hat{G}_w^S(\bar{x})$ are closed, convex and nonempty.*

We turn to a proof of theorem 9.1.6. Here is an outline of the proof: For each Bayesian society

$$\mathcal{S} := \left(\{C^j, T^j, u^j, \pi\}_{j \in N}, \{C_0^S, F^S\}_{S \in \mathcal{N}}\right) \in \text{SPACE},$$

construct the associated *auxiliary Bayesian society*,

$$\mathcal{S}_a := \left(\{C^j, T^j, u^j, \pi\}_{j \in N}, \{C_0^S, G^S\}_{S \in \mathcal{N}}\right).$$

A *Bayesian strongly incentive compatible strong equilibrium* (in short, a BSIC strong equilibrium) of $\mathcal{S}_a$ is defined as a strategy bundle $x^* \in \hat{G}_s^N$ such that it is not true that there exist $S \in \mathcal{N}$ and $u \in V_{x^*}(S)$ for which $Eu^j(x^*) < u_j$ for all $j \in S$. In view of facts 9.A.3 and 9.A.4, if x^* is a BSIC strong equilibrium of $\mathcal{S}_a$ and if x_1^{*j} is 1-1 on T^j for every $j \in N$, then x^* is a Bayesian incentive compatible strong equilibrium of the original Bayesian society $\mathcal{S}$. This motivates the following definition:

DEFINITION 9.A.6 The subset SPACE_0 of SPACE consists of all $\mathcal{S} \in$ SPACE such that there exists a BSIC strong equilibrium x^* of $\mathcal{S}_a$ for which
(i) x_1^{*j} is 1-1 on T^j for every $j \in N$, and
(ii) there exits an open neighborhood $U(Eu^*)$ of $Eu(x^*)$ in $\mathbf{R}^N$ so that $U(Eu^*) \bigcap V_{x^*}(S) = \emptyset$ for any $S \in \mathcal{N} \setminus \{N\}$.

It suffices to show that SPACE_0 is open and dense in (SPACE, d).

The following lemma 9.A.7 can be proved exactly in the same way as lower semicontinuity of $\hat{F}^S$ was proved in the proof of theorem 8.2.1:

LEMMA 9.A.7 *Under assumptions 1, 3 and 8, both $\hat{G}_s^S$ and $\hat{G}_w^S$ are lower semicontinuous in X.*

Due to lemma 9.A.7, one can establish the following lemma 9.A.8 by direct application of the social coalitional equilibrium existence theorem (theorem 8.A.5):

LEMMA 9.A.8 *If a Bayesian society $\mathcal{S}$ satisfies assumptions 1, 3, 7 and 8, then there exists a BSIC strong equilibrium in the associated auxiliary Bayesian society $\mathcal{S}_a$.*

Denote by $\hat{K}_s$ the set of all $x \in X$ such that

$$\forall j \in N : \forall \bar{t} \in T : \forall \tilde{t}^j \in T^j :$$
$$Eu^j(x \mid \bar{t})$$
$$\geq Eu^j\left((x_1^j(\tilde{t}^j), x_2^j(\tilde{t}^j, \bar{t}^{N\setminus\{j\}})),\right.$$
$$\left.(x_1^{N\setminus\{j\}}(\bar{t}^{N\setminus\{j\}}), x_2^{N\setminus\{j\}}(\tilde{t}^j, \bar{t}^{N\setminus\{j\}})) \mid \bar{t}\right).$$

The set $\hat{K}_s$ is the intersection of a convex polyhedral cone and the space X. Notice that $\hat{G}_s^N = G^N \cap \hat{K}_s$. Let $\overset{\circ}{\hat{K}}_s$ denote the interior of $\hat{K}_s$ in X.

A strategy bundle $\bar{x} \in \hat{G}_s^N$ is called a *BSIC Pareto optimum*, if there exists no $x' \in \hat{G}_s^N$ for which $Eu(x') \gg Eu(\bar{x})$. Define the unit simplex in $\mathbf{R}^N$,

$$\Delta^N := \left\{\theta \in \mathbf{R}_+^N \;\middle|\; \sum_{j \in N} \theta_j = 1\right\}.$$

Since $\hat{G}_s^N$ is convex, a strategy bundle $\bar{x}$ is BSIC Pareto optimal, iff there exists $\theta \in \Delta^N$ such that $\bar{x}$ is a solution to:

$$\text{Problem } P(\theta, \mathcal{S}): \quad \text{Maximize} \sum_{j \in N} \theta_j Eu^j(x)$$
$$\text{subject to} \quad x \in \hat{G}_s^N.$$

Proof of Theorem 9.1.6 Openness of SPACE_0: Choose any $\mathcal{S} \in \text{SPACE}_0$. Then, there exist a BSIC strong equilibrium x^* of $\mathcal{S}_a$ and a neighborhood $U(Eu^*)$ of $Eu(x^*)$ as in definition 9.A.6. One may choose a neighborhood $O(x^*)$ of x^* and $\delta > 0$ such that for each $x \in O(x^*)$, x_1^j is 1-1 on T^j, and such that for all $\mathcal{S}'$ satisfying $d(\mathcal{S}', \mathcal{S}) < \delta$ and for all $x \in O(x^*)$,

$$Eu(x) \notin \bigcup_{S \in \mathcal{N}\setminus\{N\}} V'_x(S).$$

The strategy bundle x^* is BSIC Pareto optimal; let $\theta^* \in \Delta^N$ be the associated weights, and denote by v^* the objective function of problem $P(\theta^*, \mathcal{S})$ for x^*:

$$v^*(x) := \sum_{j \in N} \theta_j^* Eu^j(x).$$

Function v^* is affinely linear, so all the contour sets of v^* are parallel hyperplanes. There are two possible cases: *Case (1)*: The contour sets of v^* are not parallel to any non-trivial proper face of $\hat{K}_s$; and *Case (2)*: The contour sets of v^* are parallel to some non-trivial proper face H^* of $\hat{K}_s$.

Consider the parameterized problems $P(\theta^*, \mathcal{S}')$, $\mathcal{S}' \in \text{SPACE}$. By definition of pseudo-metric d, the constraint set of $P(\theta^*, \mathcal{S}')$ depends upper and lower semicontinuously on $\mathcal{S}'$. By Berge's maximum theorem, the solution set $\Phi^*(\mathcal{S}')$ for $P(\theta^*, \mathcal{S}')$ depends upper semicontinuously on $\mathcal{S}'$.

Case (1): In the light of assumption 3, the solution set $\Phi^*(\mathcal{S})$ for $P(\theta^*, \mathcal{S})$ is the singleton $\{x^*\}$ in this case. Then, by the upper semicontinuity of $\Phi^*(\cdot)$, there exists $\delta_1 > 0$ such that for all $\mathcal{S}'$ for which $d(\mathcal{S}', \mathcal{S}) < \delta_1$, $\Phi^*(\mathcal{S}') \cap O(x^*) \neq \emptyset$. Therefore, for all $\mathcal{S}'$ for which $d(\mathcal{S}', \mathcal{S}) < \min\{\delta, \delta_1\}$, $\mathcal{S}' \in \text{SPACE}_0$.

Case (2): There are three subcases: *Subcase (2.1)*: $H^* \cap G^N = \emptyset$; *Subcase (2.2)*: $H^* \cap G^N = \{x^*\}$; and *Subcase (2.3)*: $H^* \cap G^N$ has more than one element.

Subcase (2.1): In this case, $x^* \in \overset{\circ}{\hat{K}}_s$. Then, by assumption 3, the solution set $\Phi^*(\mathcal{S})$ for $P(\theta^*, \mathcal{S})$ is the singleton $\{x^*\}$. The same argument as in case 1 applies.

Subcase (2.3): For all $\mathcal{S}'$ sufficiently close to $\mathcal{S}$, the set $H^* \cap G'^N$ is the solution set for $P(\theta^*, \mathcal{S}')$, and $O(x^*) \cap H^* \cap G'^N \neq \emptyset$; hence $\mathcal{S}' \in \text{SPACE}_0$.

Subcase (2.2): For any $\mathcal{S}'$, let $\Psi^*(\mathcal{S}') := \{x \in G' \mid v^*(x) = \max v^*(G')\}$. By assumption 3, $\Psi^*(\cdot)$ is a singleton, so it is continuous. Choose $\delta_2 > 0$ so that for all $\mathcal{S}'$ for which $d(\mathcal{S}', \mathcal{S}) < \delta_2$, $\Psi^*(\mathcal{S}') \in O(x^*)$. If $\Psi^*(\mathcal{S}') \in \hat{K}_s$, then $\Phi^*(\mathcal{S}') = \Psi^*(\mathcal{S}')$. If $\Psi^*(\mathcal{S}') \notin \hat{K}_s$, then $\Phi^*(\mathcal{S}') = H^* \cap G'^N$. Therefore, for all $\mathcal{S}'$ sufficiently close to $\mathcal{S}$, $\mathcal{S}' \in \text{SPACE}_0$.

Denseness of SPACE_0: Using the specific strategy bundles x_m and x_M of assumption 6, define Bayesian society $\mathcal{S}^\dagger$ by: For all $\bar{x} \in X$,

$$F^{\dagger S} = F^{\dagger S}(\bar{x}) := \{x_m^S\}, \text{ if } S \neq N,$$
$$F^{\dagger N} = F^{\dagger N}(\bar{x}) := \text{co } \{x_m, x_M\}.$$

Clearly $\mathcal{S}^\dagger \in \text{SPACE}_0$. Indeed, x_M is the unique BSIC strong equilibrium of $\mathcal{S}^\dagger$ ($= \mathcal{S}_a^\dagger$), and satisfies conditions (i) and (ii) of definition 9.A.6.

Choose any $\mathcal{S} \in \text{SPACE}$. By lemma 9.A.8, the auxiliary Bayesian society $\mathcal{S}_a$ possesses a BSIC strong equilibrium x^*. For any neighborhood $U(\mathcal{S})$ of $\mathcal{S}$, one needs to construct a Bayesian society in $U(\mathcal{S}) \cap \text{SPACE}_0$. Define for each $\alpha \in [0, 1]$, the Bayesian society $\mathcal{S}^\alpha$ by:

$$F^{\alpha S}(\alpha \bar{x} + (1-\alpha) x_M) := \alpha F^S(\bar{x}) + (1-\alpha) F^{\dagger S}.$$

Define

$$x^{\alpha *}(\cdot) := \alpha x^*(\cdot) + (1-\alpha) x_M(\cdot).$$

Due to the affine linearity of $u^j(\,\cdot\,,t)$, for all $\alpha \in [0,1)$, $x^{\alpha*}$ is a BSIC strong equilibrium of $\mathcal{S}_a^\alpha$, and condition (ii) of definition 9.A.6 is satisfied for this $x^{\alpha*}$. Since $x_{M,1}^j$ is 1-1 on T^j, so is $x_1^{\alpha*j}$ for all α sufficiently close to 1. □

The third purpose of this appendix is to prove theorem 9.2.7. One of Scarf's results on increasing returns to scale is recalled first: Consider a static coalition production economy with one primary good and one final good,

$$(\{\Sigma^j,\ u^j,\ r^j\}_{j\in N},\ c)\,,$$

where N is a finite set of economic agents, Σ^j ($\subset \mathbf{R}_+^2$) is a strategy set of agent j ($\sigma^j := (\eta^j,\zeta^j) \in \Sigma^j$ means j's consumption of η^j units of the final good and ζ^j units of the primary good), $u^j : \Sigma^j \to \mathbf{R}$ is a utility function of agent j, $r^j > 0$ is the initial resource of primary good held by agent j, and $c : \mathbf{R}_+ \to \mathbf{R}_+$ is the cost function, which is common to all coalitions ($c(\eta)$ is the minimal amount of the primary good that is needed to produce η units of the final good).

THEOREM 9.A.9 (Scarf, 1973) *Let $(\{\Sigma^j,\ u^j,\ r^j\}_{j\in N},\ c)$ be a coalition production economy with one primary good and one final good. Define a non-side-payment game $V : \mathcal{N} \to \mathbf{R}^N$ by*

$$V(S) := \left\{ u \in \mathbf{R}^N \;\middle|\; \begin{array}{l} \exists\,(\eta^j,\zeta^j)_{j\in S} \in \prod_{j\in S}\Sigma^j : \\ c\left(\sum_{j\in S}\eta^j\right) + \sum_{j\in S}\zeta^j \leq \sum_{j\in S}r^j, \\ \forall\,i \in S : u_i \leq u^i(\eta^i,\zeta^i) \end{array} \right\}.$$

If each u^j is nondecreasing in $\mathbf{R}_+^2$, and if

$$0 < \eta < \eta' \implies \frac{c(\eta)}{\eta} \geq \frac{c(\eta')}{\eta'},$$

then the game V is balanced.

Theorem 9.2.7 will be proved by applying the technique of Ichiishi and Quinzii (1983) modified suitably for the present purpose: Roughly stated, a family of societies $\Gamma := (\{\Sigma^j, u^j\}_{j\in N},\ \{\Phi^S\}_{S\in\mathcal{N}},\ Q,\ \Psi,\ \{\{N\}\})$ parameterized by price vector $q \in Q$ (definition 8.A.7) is constructed from the profit-center game

$$(\{\mathbf{R}^{k_m+k_n}, T^j,\ \text{profit function}, r^j, \pi\}_{j\in N}, \{Y^j\}_{j\in N}, p)\,.$$

The parameterized family Γ will be shown to satisfy the assumptions of the social coalitional equilibrium existence theorem 8.A.9. From a social coalitional equilibrium of Γ, one can easily construct the required core plan of the original profit-center game. Recall, however, the cost-minimization problem $P^N(q, \eta)$ and the partition of the price domain into $\{Q_+, Q_0\}$, defined in section 9.2. To be precise, in view of the fact that problem $P^N(q, \eta)$ may not have an optimal solution if $q \in Q_0$, a *sequence* of parameterized families of societies $\{\Gamma^\nu\}_{\nu=1}^\infty$, $Q^\nu \subset Q_+$, is actually constructed, a social coalitional equilibrium $(q^\nu, \sigma^{\nu,N})$ of Γ^ν is chosen for each ν, and a limit point $(\bar{q}, \bar{\sigma}^N)$ of the sequence $\{q^\nu, \sigma^{\nu,N}\}_\nu$ is considered.

The definition of Γ^ν is now presented. The fictitious price vector q for the type-profile-contingent nonmarketed commodities, introduced in the second paragraph following theorem 9.2.5, will play the role of parameter in Γ^ν. Define

$$Q^\nu := \left\{ q \in Q \,\middle|\, \forall\, t \in T : \forall\, a \in K_n : q(t,a) \geq \frac{1}{\nu} \right\}.$$

Since $Q^\nu \subset Q_+$, problem $P^N(q,\eta)$ has an optimal solution for all $(q,\eta) \in Q^\nu \times \overset{\circ}{\mathbf{R}}_+$. The parameter space of Γ^ν is, therefore, the trimmed simplex Q^ν.

Given the resource constraint r^N, define a lower bound and an upper bound for the maximal profit:

$$\underline{\eta} := \min_{j \in N} \max \left\{ \min_{t \in T} \{p \cdot y_m^j(t)\} \,\middle|\, \begin{pmatrix} y_m^j \\ -r^j \end{pmatrix} \in Y^j \atop y_{1m}^j \text{ is } \mathcal{T}^j\text{-measurable} \right\},$$

$$\bar{\eta} := \max \left\{ \max_{t \in T} \{p \cdot \sum_{j \in N} y_m^j(t)\} \,\middle|\, \begin{pmatrix} y_m^N \\ -r^N \end{pmatrix} \in Y^N \atop y_{1m}^j \text{ is } \mathcal{T}^j\text{-measurable} \right\}.$$

¿From assumption 9.2.3 (v) and the present conditions ((1) positive profitability from strict inputs and (2) strict positiveness of $r^j(t^j)$), it follows that $0 < \underline{\eta} \leq \bar{\eta} < \infty$. The strategy spaces are defined as

$$\Sigma^j := [\underline{\eta}, \bar{\eta}] \times \{0\} \, (\subset \overset{\circ}{\mathbf{R}}_+ \times \mathbf{R}_+),\ j \in N.$$

Division j's strategy $(\eta^j, \zeta^j) \in \Sigma^j$ is identified with $\eta^j \in [\underline{\eta}, \bar{\eta}]$ since $\zeta^j \equiv 0$, and is interpreted as the profit attributed to j. To use the language of Scarf's model with one primary good and one final good (see theorem 9.A.9), (η^j, ζ^j) means j's consumption of η^j units of the "final good", and no consumption of the "primary good". Define $\Sigma^S := \prod_{i \in S} \Sigma^i$, and set $\Sigma := \Sigma^N$.

The feasible-strategy correspondences $\Phi^S : Q \to \Sigma^S$, $S \in \mathcal{N}$, are defined

as
$$\Phi^S(\bar{q}) := \left\{ (\eta^S, 0) \in \Sigma^S \,\middle|\, \hat{c}_w^S\left(\bar{q}, \sum_{i \in S} \eta^i\right) \leq \sum_{i \in S}\sum_{t \in T} \bar{q}(t) \cdot r^i(t^i) \right\}.$$

(To be precise, the restriction of Φ^S to Q^ν is used for Γ^ν.)

The utility functions of Γ^ν are given as
$$u^j(q, \eta^j, \zeta^j) := \eta^j, \; j \in N.$$

LEMMA 9.A.10 *There exists a continuous function*
$$\begin{aligned}(x^{*N}, y^*): \; & Q_+ \times \overset{\circ}{\mathbf{R}}_+ \;\to\; \mathbf{R}^{|T \times N|} \times \mathbf{R}^{k|T|}, \\ & (q, \eta) \;\mapsto\; (x^{*N}(q,\eta)(\cdot), \, y^*(q,\eta)(\cdot))\end{aligned}$$
such that for each (q, η), $(x^{*N}(q,\eta), y^*(q,\eta)) : T \to \mathbf{R}^{|N|} \times \mathbf{R}^k$ is an optimal solution to problem $P^N(q, \eta)$.

Proof Step 1. Given $q \in Q_+$ and any function $x : T \to \mathbf{R}$, the problem,
$$\text{Minimize} \quad \sum_{t \in T} q(t) \cdot (-y_n(t)),$$
$$\text{subject to} \quad -y_n \in S_{Y^N}(y_m), \text{ and}$$
$$\forall \, t: \; x(t) \leq p \cdot y_m(t),$$

has a unique solution $y^*(q, x)$. Indeed, let $y := (y_m, y_n)$ and $y' := (y'_m, y'_n)$ be solutions. Then,
$$g(q, y_m) = g(q, y'_m).$$
If $y_m \neq y'_m$, set $y''_m := (y_m + y'_m)/2$. By strict quasi-convexity of $g(q, \cdot)$,
$$g(q, y''_m) < g(q, y_m),$$
and clearly
$$\forall \, t \in T: \; x(t) \leq p \cdot y''_m(t),$$
which contradicts the optimality of y. Therefore $y_m = y'_m$. By strict convexity of $S_{Y^N}(y_m)$, $y_n = y'_n$.

Step 2. Using the subvector $y_n^*(q, x)$ of $y^*(q, x)$ obtained for an arbitrary $x \in \mathbf{R}^T$ in step 1, problem $P^N(q, \eta)$ is equivalent to finding a solution $x^N : T \to \mathbf{R}^N$ to:
$$\text{Minimize} \quad -\sum_{t \in T} q(t) \cdot y_n^*\left(q, \sum_{i \in N} x^i\right)(t),$$
$$\text{subject to} \quad \eta \leq \sum_{i \in N} \min_{t \in T} E(x^i \mid \mathcal{T}^i)(t).$$

Let x^N and x'^N be two optimal solutions of this problem. Then, one may assume without loss of generality,

$$\forall\, t:\ \sum_{i\in N} x^i(t) = \sum_{i\in N} x'^i(t) =: x^*(q,\eta)(t).$$

To show this, notice first that

$$y_m^*\left(q, \sum_{i\in N} x^i\right) = y_m^*\left(q, \sum_{i\in N} x'^i\right).$$

Indeed, if this last equality is false, set

$$x''^N := \frac{x^N + x'^N}{2},$$

$$y_m'' := \frac{y_m^*(q, \sum_{i\in N} x^i) + y_m^*(q, \sum_{i\in N} x'^i)}{2}.$$

Then,

$$\forall\, t:\ \sum_{i\in N} x''^i(t) \le p\cdot y_m''(t),$$

$$\eta \le \sum_{i\in N} \min_{t\in T} E(x''^i \mid T^i)(t),$$

$$g(q, y_m'') < g\left(q, y_m^*\left(q, \sum_{i\in N} x^i\right)\right),$$

which contradicts the optimality of $(x^N, y_m^*(q, \sum_{i\in N} x^i))$. One can then define

$$\forall\, t:\ x^*(q,\eta)(t) := p\cdot y_m^*\left(q, \sum_{i\in N} x^i\right)(t).$$

Step 3. By the maximum theorem, the function $(q,\eta) \mapsto x^*(q,\eta)(\cdot)$ is continuous on $Q_+\times \mathring{\mathbf{R}}_+$. The set of optimal solutions of the problem $P^N(q,\eta)$ is now given as

$$\left\{ x^N : T \to \mathbf{R}^N \ \middle|\ \begin{array}{l} \forall\, t\in T: \\ \sum_{i\in N} E(x^i \mid t^i) \ge \eta, \\ \sum_{i\in N} x^i(t) \le x^*(q,\eta)(t) \end{array} \right\}.$$

By Walkup and Wets (1969), this polyhedron depends continuous piecewise linearly on the right-hand-side parameter of the constraint, $(\eta, x^*(q,\eta)(\cdot))$;

that is, there are finitely many continuous piecewise linear functions h^j, $j \in J$, such that for each $(\eta, x^*(q, \eta)(\cdot))$,

$$\{\text{the extreme points of the polyhedron given } (\eta, x^*(q, \eta)(\cdot))\} \subset \{h^j(\eta, x^*(q, \eta)(\cdot)) \mid j \in J\}.$$

In particular, each extreme point depends continuously on (q, η). This establishes the required choice of $x^{*N} : Q_+ \times \overset{\circ}{\mathbf{R}}_+ \to \mathbf{R}^{|T \times N|}$.

Step 4. Define

$$y^*(q, \eta)(\cdot) := y^*(q, x^*(q, \eta)(\cdot))(\cdot),$$

where y^* of the right-hand side is given in step 1 and x^* is given in step 2.
□

The parameter's response correspondence $\Psi^\nu : Q^\nu \times \Sigma \to Q^\nu$ of the society Γ^ν is defined by: $q^\circ \in \Psi^\nu(\bar{q}, (\bar{\eta}^N, \mathbf{0}))$ iff q° maximizes the value of the "total excess demand for the 'nonmarketed commodities'". That is, q° solves the mathematical programming problem,

Maximize $\displaystyle\sum_{t \in T} q(t) \cdot \left(-y_n^* \left(\bar{q}, \sum_{j \in N} \bar{\eta}^j \right)(t) - \sum_{i \in N} r^i(t^i) \right),$

subject to $\quad q \in Q^\nu$,

where $y_n^*(\bar{q}, \sum_{j \in N} \bar{\eta}^j)(\cdot) : T \to \mathbf{R}^{k_n}$ is given in lemma 9.A.10. Solution q° $(\in \Psi(\bar{q}, (\bar{\eta}^N, \mathbf{0})))$ assigns a large weight (*i.e.*, $q^\circ(\bar{t}, \bar{a}) > 1/\nu$) only if the excess demand for $(\bar{t}, \bar{a})$ is the greatest for all $(t, a) \in T \times K_n$.

Proof of Theorem 9.2.7 Step 1. For each ν large enough so that $Q^\nu \neq \emptyset$, the family of parameterized societies

$$\Gamma^\nu := (\{\Sigma^i\}_{i \in N}, Q^\nu, \{\Phi^S\}_{S \in \mathcal{N}}, \Psi^\nu, \{u^i\}_{i \in N})$$

that is constructed from the profit-center game

$$(\{\mathbf{R}^{k_m + k_n}, T^j, \text{profit function}, r^j, \pi\}_{j \in N}, \{Y^j\}_{j \in N}, p)$$

is well-defined and satisfies conditions (i) and (ii) of theorem 8.A.9.

One may assume without loss of generality that $\hat{c}_w^S = \hat{c}_w^N$ for all S. Indeed, by assumption 9.2.3 (iii), $\hat{c}_w^S \geq \hat{c}_w^N$ for all S. A core plan of the (hypothetical) profit-center game in which every coalition S has access to the cost function $\hat{c}_w^N$ is also a core plan of the (original) profit-center game, since in the latter (original) game the "blocking power" of S is weaker, so more plans are coalitionally stable than in the former (hypothetical) game.

The family Γ^ν satisfies condition (i') (iii) of theorem 8.A.9. Indeed, in view of assumption 9.2.6 (iii) and strict positiveness of $r^j(t^j)$, the standard technique in the neoclassical consumer theory to establish upper and lower semicontinuity of the budget-set correspondence is applicable.

The family Γ^ν satisfies condition (iv) of theorem 8.A.9. Indeed, by assumption 9.2.6 (i), for each given $q \in Q$,

$$0 < \eta < \eta' \Longrightarrow \frac{\hat{c}_w^N(q,\eta)}{\eta} \geq \frac{\hat{c}_w^N(q,\eta')}{\eta'}.$$

Therefore, for each $q \in Q$, the non-side-payment game $\hat{V}_{w,q} : \mathcal{N} \to \mathbf{R}^N$ defined by

$$\hat{V}_{w,q}(S) := \{u \in \mathbf{R}^N \mid \exists\, \sigma^S \in \Phi^S(q) : \forall\, j \in S : u^j \leq u^j(q, \sigma^j)\}$$

is balanced by theorem 9.A.9.

The family Γ^ν satisfies condition (v) of theorem 8.A.9. Indeed, choose any $\bar{q} \in Q^\nu$ and any utility allocation $\bar{u} \in \mathbf{R}^N$. Let $\sigma^{0N}, \sigma^{1N} \in \Phi^N(\bar{q})$ be such that

$$\forall\, j \in N : \bar{u}_j \leq \min\left\{u^j(\bar{q}, \sigma^{0j}), u^j(\bar{q}, \sigma^{1j})\right\},$$

and for each $\alpha \in [0,1]$ define $\sigma^{\alpha N} \in \Sigma$ by

$$\sigma^{\alpha N} := \alpha \sigma^{1N} + (1-\alpha)\sigma^{0N}.$$

Clearly, $\bar{u}_j \leq u^j(\bar{q}, \sigma^{\alpha j})$ for all $j \in N$. It suffices to show that $\sigma^{\alpha N} \in \Phi^N(\bar{q})$. Let $(x^{0N}, y^0) \in C^N(\sum_{i \in N} \eta^{0i})$ be an optimal solution to problem $P^N(\bar{q}, \sum_{i \in N} \eta^{0i})$. Then,

$$g(\bar{q}, y_m^0) = \hat{c}_w^N\left(\bar{q}, \sum_{i \in N} \eta^{0i}\right) \leq \sum_{i \in N} \sum_{t \in T} \bar{q}(t) \cdot r^i(t^i).$$

Similarly, for an optimal solution (x^{1N}, y^1) to problem $P^N(\bar{q}, \sum_{i \in N} \eta^{1i})$,

$$g(\bar{q}, y_m^1) \leq \sum_{i \in N} \sum_{t \in T} \bar{q}(t) \cdot r^i(t^i).$$

Define
$$(x^{\alpha N}, y_m^\alpha) := \alpha(x^{1N}, y_m^1) + (1-\alpha)(x^{0N}, y_m^0).$$

By assumption 9.2.6 (iv), for any $\alpha \in [0,1]$,

$$g(\bar{q}, y_m^\alpha) \leq \sum_{i \in N} \sum_{t \in T} \bar{q}(t) \cdot r^i(t^i).$$

On the other hand, for any y_n^j for which y_{1n}^j is T^j-measurable and $(y_m^\alpha, \sum_{i\in N} y_n^i) \in \sum_{i\in N} Y^i$, it follows that $(x^{\alpha N}, y_m^\alpha, \sum_{i\in N} y_n^i) \in C^N(\sum_{i\in N} \eta^{\alpha i})$, so that

$$\hat{c}_w^N\left(\bar{q}, \sum_{i\in N} \eta^{\alpha i}\right) \leq -\sum_{i\in N}\sum_{t\in T} \bar{q}(t) \cdot y_n^i(t),$$

and consequently,

$$\hat{c}_w^N\left(\bar{q}, \sum_{i\in N} \eta^{\alpha i}\right) \leq g(\bar{q}, y_m^\alpha).$$

Therefore,

$$\hat{c}_w^N\left(\bar{q}, \sum_{i\in N} \eta^{\alpha i}\right) \leq \sum_{i\in N}\sum_{t\in T} \bar{q}(t) \cdot r^i(t^i),$$

that is, $\sigma^{\alpha N} \in \Phi(\bar{q})$.

The family Γ^ν satisfies condition (iii$'$) of theorem 8.A.9 by the maximum theorem.

Thus, all the assumptions of theorem 8.A.9 are satisfied. Let $(q^\nu, \sigma^{\nu N}) \in Q^\nu \times \Sigma$ be a social coalitional equilibrium of Γ^ν. By definition of Ψ^ν, $q^\nu \in Q_+$. Let

$$(x^{\nu N}, y^\nu) := \left(x^{*N}\left(q^\nu, \sum_{i\in N} \eta^{\nu i}\right), y^*\left(q^\nu, \sum_{i\in N} \eta^{\nu i}\right)\right);$$

it is an optimal solution to problem $P^N(q^\nu, \sum_{i\in N} \eta^{\nu i})$. Let $y^{\nu N} \in Y^N$ give rise to y^ν (in particular, $y^\nu = \sum_{i\in N} y^{\nu i}$). One may choose $y^{\nu N}$ so that each $y_1^{\nu j}$ is T^j-measurable.

Step 2. The sequence $\{y_n^\nu\}_\nu$ ($\subset -\mathbf{R}_+^{k_2|T|}$) obtained in step 1 is bounded. To show this, suppose the contrary. Then, there exists a subsequence, still denoted by $\{y_n^\nu\}_\nu$, such that

$$y_n^\nu(t^\nu, a^\nu) \to -\infty, \quad \text{as} \quad \nu \to \infty, \tag{9.7}$$

for some choice of coordinate (t^ν, a^ν) for each ν. For each ν, define L^ν ($\subset T \times K_n$) as the set of (type-profile, commodity)-pairs whose excess demand is maximal. That is, $(\bar{t}, \bar{a}) \in L^\nu$ iff

$$-y_n^\nu(\bar{t}, \bar{a}) - \sum_{i\in N} r^i(\bar{t}^i, \bar{a}) = \max_{(t,a)}\left\{-y_n^\nu(t, a) - \sum_{i\in N} r^i(t^i, a)\right\}.$$

By passing through a subsequence if necessary, one may assume without loss of generality that

$$L^\nu = L^{\nu+1} = \cdots =: L^\circ.$$

By the present hypothesis (9.7),

$$\forall\, (t,a) \in L^\circ : y_n^\nu(t,a) \to -\infty, \text{ as } \nu \to \infty.$$

This holds true, only if $(q^\nu, \sigma^{\nu N})$ as a point in the domain $Q^\nu \times \Sigma$ of the correspondence Ψ^ν satisfies

$$\forall\, (t,a) \in L^\circ : q^\nu(t,a) \to 0, \text{ as } \nu \to \infty$$

(otherwise, the cost $-\sum_{t \in T} q^\nu(t) \cdot y_n^\nu(t)$ would be arbitrarily large, which contradicts the definition of y_n^ν as a cost-minimizer). Therefore,

$$\sum_{(t,a) \in L^\circ} q^\nu(t,a) \to 0, \quad \text{as} \quad \nu \to \infty. \tag{9.8}$$

On the other hand, q^ν as a point in the response set $\Psi^\nu(q^\nu, \sigma^{\nu N})$ assigns a greater weight than $1/\nu$ only to members of L°. So

$$\sum_{(t,a) \in L^\circ} q^\nu(t,a) \;\geq\; 1 - \frac{k_n |T|}{\nu}$$

$$\to\; 1, \text{ as } \nu \to \infty,$$

which contradicts (9.8).

Step 3. Since $\{y_n^\nu\}_\nu$ is bounded, and since

$$y_n^\nu = \sum_{i \in N} y_n^{\nu i},$$

$$y_n^{\nu i} \leq \mathbf{0} \text{ for every } i \in N,$$

the sequence $\{y_n^{\nu i}\}_\nu$ is also bounded for each i. By assumption 9.2.3 (v), $\{y_m^{\nu i}\}_\nu$ is bounded from above, and hence so is $\{y_m^\nu\}_\nu$. Moreover,

$$\forall\, t \in T : \eta^\nu \leq p \cdot E(y_m^\nu \mid \mathcal{T}^i)(t),$$

so $\{y_m^\nu\}_\nu$ is bounded from below as well. Therefore, $\{y_m^{\nu j}\}_\nu$ is also bounded. Since $(x^{\nu N}, y^\nu) \in C^N(\eta^\nu)$, $\{x^{\nu N}\}_\nu$ is bounded. Thus, one may assume without loss of generality,

$$\begin{aligned}
q^\nu &\to \bar{q} \in Q, \\
\sigma^{\nu N} &\to \bar{\sigma}^N \in \Sigma, \\
x^{\nu N} &\to \bar{x}^N \in \mathbf{R}^{|T \times N|}, \\
y^{\nu N} &\to \bar{y}^N \in Y^N, \text{ as } \nu \to \infty.
\end{aligned}$$

9: Information Revelation

Step 4. The plan $(\bar{x}^N, \bar{y}^N)$ is a member of F^N. To show this, one only needs to check

$$\forall\, t \in T: \quad -\sum_{i \in N} \bar{y}^i_n(t) \leq \sum_{i \in N} r^i(t^i). \tag{9.9}$$

For this purpose, choose any q in the relative interior of Q. Then, $q \in Q^\nu$ for all ν sufficiently large, so that

$$\begin{aligned}
\sum_{t \in T} q(t) \cdot \left(-\sum_{i \in N} y^{\nu i}_n(t) - \sum_{i \in N} r^i(t^i) \right) \\
\leq \sum_{t \in T} q^\nu(t) \cdot \left(-\sum_{i \in N} y^{\nu i}_n(t) - \sum_{i \in N} r^i(t^i) \right) \\
= \hat{c}^N_w\left(q^\nu, \sum_{i \in N} \eta^{\nu i} \right) - \sum_{i \in N} \sum_{t \in T} q^\nu(t) \cdot r^i(t^i) \\
\leq 0.
\end{aligned}$$

Letting $\nu \to \infty$,

$$\sum_{t \in T} q(t) \cdot \left(-\sum_{i \in N} \bar{y}^i_n(t) - \sum_{i \in N} r^i(t^i) \right) \leq 0.$$

This is true for all q in the relative interior of Q, hence it is also true for all $q \in Q$. In particular, it is true for $q = e^{\tilde{t},\tilde{a}} \in Q$ defined by

$$e^{\tilde{t},\tilde{a}}(t,a) := \begin{cases} 1, & \text{if } (t,a) = (\tilde{t}, \tilde{a}), \\ 0, & \text{otherwise.} \end{cases}$$

This means that

$$-\sum_{i \in N} \bar{y}^i_n(\tilde{t}, \tilde{a}) - \sum_{i \in N} r^i(\tilde{t}^i, \tilde{a}) \leq 0.$$

Since $(\tilde{t}, \tilde{a})$ was chosen arbitrarily, (9.9) is now established.

Since $\eta^{\nu N}$ $(:= (u^i(q^\nu, \sigma^{\nu i}))_{i \in N})$ is in the core of game $\hat{V}_{w,q^\nu}$ for every ν, and since Φ^S's are upper and lower semicontinuous, it follows that $\bar{\eta}^N$ is in the core of game $\hat{V}_{w,\bar{q}}$:

$$\bar{\eta}^N \in \hat{V}_{w,\bar{q}}(N), \tag{9.10}$$

and

$$\neg \exists\, S \in \mathcal{N}: \exists\, \eta^N \in \hat{V}_{w,\bar{q}}(S): \forall\, j \in S: \eta^j > \bar{\eta}^j. \tag{9.11}$$

By (9.10), $(\bar{x}^N, \bar{y}^N) \in C^N(\sum_{i \in N} \bar{\eta}^i)$. Define a profit-imputation plan $\bar{x}'^N$ by:
$$\bar{x}'^j(t^j, t^{N\setminus\{j\}}) := \bar{\eta}^j - E(\bar{x}^j \mid T^j) + \bar{x}^j(t^j, t^{N\setminus\{j\}}).$$
Then,
$$\begin{aligned} E(\bar{x}'^j \mid T^j)(t) &= \bar{\eta}^j - E(\bar{x}^j \mid T^j)(t) + E(\bar{x}^j \mid T^j)(t) \\ &= \bar{\eta}^j, \end{aligned}$$
which is independent of t. Moreover, for all $t \in T$,
$$\begin{aligned} \sum_{i \in N} \bar{x}'^i(t) &= \sum_{i \in N} \bar{\eta}^i - \sum_{i \in N} E(\bar{x}^i \mid T^i)(t) + \sum_{i \in N} \bar{x}^i(t) \\ &\leq \sum_{i \in N} \bar{x}^i(t) \\ &\leq \sum_{i \in N} p \cdot \bar{y}_m^i(t). \end{aligned}$$

Therefore, $(\bar{x}'^N, \bar{y}^N) \in \hat{G}_w^N$. Define $\bar{\bar{x}}^j := E(\bar{x}'^j \mid T^j)$. Then, $\bar{\bar{x}}^j(t) \equiv \bar{\eta}^j$. By (9.11) no coalition S can improve upon $(\bar{\bar{x}}^N, \bar{y}^N)$ using strategies in $\hat{G}_w^S$.

Using lemma 9.2.10 and postulate 9.2.1, one can construct the required full-information revealing, Bayesian incentive-compatible core plan from $(\bar{x}'^N, \bar{y}^N)$, as in the last two paragraphs of the proof of theorem 9.2.5. □

Part III

PURE EXCHANGE ECONOMY

Chapter 10

Existence

To date many works have been done on the Bayesian pure exchange economy (example 2.2.1) with l commodities,

$$\mathcal{E}_{pe} := \left\{ C^j, T^j, u^j, e^j, \{\pi^j(\cdot \mid t^j)\}_{t^j \in T^j} \right\}_{j \in N},$$

where N is a finite set of consumers, and for each consumer j, C^j is his consumption set, T^j is his finite type set, $u^j : C^j \times T \to \mathbf{R}$ is his type-profile dependent von Neumann-Morgenstern utility function, $e^j : T^j \to \mathbf{R}_+^l$ is his initial endowment vector, which depends only upon t^j, and $\pi^j(\cdot \mid t^j)$ is a conditional probability on $T^{N \setminus \{j\}}$ given t^j, objective or subjective. These works are mostly on the existence of a core allocation, and on the core convergence theorem. We assume that either each consumption set is the nonnegative orthant $\mathbf{R}_+^l$ of the commodity space, or it is a nonempty and compact subset of $\mathbf{R}^l$ (the latter assumption can be made without loss of generality).

Actually, we have already seen some existence results for the core of the Bayesian pure exchange economy in chapter 8, as special cases of strong equilibrium existence theorems for a more general model of Bayesian society. We will present further works on the existence of an *interim* core allocation in section 10.1: Ichiishi and Yamazaki's (2004) nonemptiness result on the Bayesian incentive-compatible coarse core, Vohra's (1999) works on the Bayesian incentive-compatible coarse core for the mediator-based approach, Vohra's (1999) example of a market for a lemon whose *interim* Bayesian incentive-compatible core is empty, Ichiishi and Yamazaki's (2004) condition for nonemptiness of the *interim* Bayesian incentive-compatible core. In all these works, consumer's choice is defined as his net trade. It is unlikely that the positive results on *interim* core concepts can be extended beyond the Bayesian pure exchange economy. In section 10.2, we will first present for completeness applications of the existence result (theorem 8.2.1) to an *ex ante* core allocation, in which consumer's choice is defined as his commodity bundle, and then establish a theorem for the *ex ante* core in which consumer's choice is his net trade.

10.1 *Interim* Solutions

We present a positive existence result on the Bayesian incentive-compatible coarse core first. Here, a strategy is a net-trade plan $z^j : T \to \mathbf{R}^l$ which is individually feasible, that is, $z^j(t) + e^j(t^j) \geq \mathbf{0}$ for all $t \in T$. A function $f : \mathbf{R}^l_+ \to \mathbf{R}$ is called *weakly monotone*, if

$$[c, c' \in \mathbf{R}^l_+,\ c \leq c'] \Rightarrow f(c) \leq f(c').$$

PROPOSITION 10.1.1 (Ichiishi and Yamazaki, 2004) *Let $\mathcal{E}_{pe}$ be a Bayesian pure exchange economy, in which each player j's strategy is a T^j-measurable net trade plan z^j. Assume that $T(S) = T(S')$ for all coalitions S and S'. Assume also for each consumer j that $C^j = \mathbf{R}^l_+$, and his von Neumann-Morgenstern utility function $u^j(\cdot, t)$ is continuous, concave, and weakly monotone in $\mathbf{R}^l_+$ for every $t \in T$. Then there exists a Bayesian incentive-compatible coarse core net-trade plan.*

Proof. As in Wilson (1978), define an agent as a pair of a consumer and his private information, and define for any $S \in \mathcal{N}$ and $E \in \bigwedge_{j \in S} \left(T^j \bigcap T(S) \right)$ the admissible coalition of agents,

$$(S, E) := \left\{ (j, t^j) \in S \times T^j \ \middle|\ \emptyset \neq \left(\{t^j\} \times T^{N \setminus \{j\}} \right) \bigcap T(S) \subset E \right\}$$

(see the paragraphs at the outset of section 8.1, preceding proposition 8.1.1). Consider the game in which strategies are private measurable:

$$V'(S, E)$$
$$:= \left\{ u \in \mathbf{R}^{\sum_{j \in N} \#T^j} \ \middle|\ \begin{array}{l} \exists\ z^S \in F'^S - \{e^S\} : \\ \forall\ j \in S : (\forall\ t^j : \{t^j\} \times T^{N \setminus \{j\}} \subset E) : \\ u_{(j, t^j)} \leq E u^j(z^j + e^j \mid t^j) \end{array} \right\}.$$

Scarf's core-nonemptiness theorem (theorem 8.A.2 of this book) is applicable to this game. Let $z^\dagger$ be a strategy bundle of the grand coalition (N, T) which gives rise to a member of the core of V'. By Ichiishi and Radner's lemma (lemma 9.2.10), there exists a private measurable strategy bundle z^* such that $z^\dagger \leq z^*$ and

$$\forall\ t \in T : \sum_{j \in N} z^*(t^j) = \mathbf{0}.$$

10: Existence

By weak monotonicity of $Eu^j(\cdot \mid t^j)$, the bundle z^* also gives rise to a member of the core of V'. By Hahn and Yannelis' proposition (proposition 4.1.3 of this book), it is Bayesian incentive-compatible. The bundle z^* is the required Bayesian incentive-compatible coarse core strategy bundle of $\mathcal{E}_{pe}$. □

The above proof is essentially a reproduction of Wilson's (1978) proof of nonemptiness of the coarse core. It shows that his technique works even when we impose the measurability condition and the Bayesian incentive compatibility condition, although Wilson did not impose either one. The idea for this positive result may be summarized as the following three steps: (1) Apply Scarf's theorem (theorem 8.A.2) to the appropriate game, such as the one with admissible coalitions of agents, in which feasible strategies are restricted to private measurable net-trade plans, and obtain a core plan. (2) Apply Ichiishi and Radner's lemma (lemma 9.2.10), and obtain another private measurable core plan for which the total demand is met by the total supply with strict equality for all type profiles. (3) Apply Hahn and Yannelis' proposition (proposition 4.1.3), and show Bayesian incentive compatibility of the core plan. Thus, the proof utilizes the specific structure of the Bayesian pure exchange economy $\mathcal{E}_{pe}$ in the private measurable case in which each consumer's strategy is his net trade plan. We have seen in the proof of a existence theorem for the general Bayesian society (theorem 8.2.1) how Bayesian incentive compatibility destroys convexity of the relevant data, even when we start out with the convex world (this was indeed the reason for the restrictive affine linearity assumption on the von Neumann-Morgenstern utility functions $u^j(\cdot, t)$ in theorem 8.2.1). The specific structure of $\mathcal{E}_{pe}$ overcomes this difficulty, so there are many positive existence results for $\mathcal{E}_{pe}$. The idea summarized in this paragraph will repeatedly show up throughout this chapter.

REMARK 10.1.2 Vohra has an example of an empty Bayesian incentive-compatible coarse core (Vohra, 1999, example 3.2, pp. 136-138), but this is within the problematic framework of mediator-based approach: It is based on his postulate that a net-trade plan of consumer j in coalition S is $\mathcal{T}^S$-measurable, rather than private-measurable. □

Vohra (1999, proposition 3.1) provided a sufficient condition for nonemptiness of a Bayesian incentive-compatible coarse core of the Bayesian pure exchange economy $\mathcal{E}_{pe}$ for the mediator-based approach. In his framework, player j's strategy is his net trade plan. Recall the notation,

$$T(\pi^j) := \bigcup_{t^j \in T^j} \left(\{t^j\} \times \operatorname{supp} \pi^j(\cdot \mid t^j) \right).$$

Information is called *non-exclusive*, if any player's unilateral deception can be detected by the other players when these other players pool information, that is, if

$$\forall j \in N : \forall\, t \in T(\pi^j) : \bigcap_{i \in N \setminus \{j\}} \{(s^{N \setminus \{i\}}, t^i) \in T \mid \pi^i(s^{N \setminus \{i\}} \mid t^i) > 0\} = \{t\}.$$

Thus the non-exclusiveness of information implies that the mediator can detect any unilateral deception. This implication is re-stated as follows: Vohra assumed that

$$\forall\, i, j \in N : \qquad T(\pi^i) = T(\pi^j). \tag{10.1}$$

Set $T^* := T(\pi^j)$. Under condition (10.1), non-exclusiveness of information is equivalent to:

$$\forall\, t \in T^* : \forall\, i, j \in N : \forall\, s^j \in T^j \setminus \{t^j\} : \pi^i(s^j, t^{N \setminus \{i,j\}} \mid t^i) = 0.$$

So, whenever player j tries to deviate from t^j to s^j ($\neq t^j$), everybody else sees that the resulting type profile $(s^j, t^{N \setminus \{j\}})$ could not have occurred. Defection is thus detected. Let z be an *interim* individually rational net-trade plan. The players can design the following alternative plan: Give the penalty of no-trade at any type profile outside the support of the *interim* probabilities. In a nutshell, unilateral deviation results in a type-profile outside the support of the *interim* probabilities, so this alternative plan is attainable and Baysian incentive-compatible, and gives rise to the same *interim* utilities as z. This is the heart of Vohra's positive result (proposition 10.1.3), and is formally presented as lemma 10.1.4.

PROPOSITION 10.1.3 (Vohra, 1999) *Let $\mathcal{E}_{pe}$ be a Bayesian pure exchange economy, in which each player j's strategy in coalition S ($\ni j$) is a T^S-measurable net trade plan z^j. Assume that (10.1) is satisfied, and that information is non-exclusive. Assume also that for each consumer j, his von Neumann-Morgenstern utility function $u^j(\cdot, t)$ is continuous, concave, and weakly monotone in $\mathbf{R}_+^l$ for every $t \in T$. Then there exists a Bayesian incentive-compatible coarse core net-trade plan.*

A key step in the proof of proposition 10.1.3 is provided by the following lemma:

LEMMA 10.1.4 (Vohra 1999) *Let $\mathcal{E}_{pe}$ be the Bayesian pure exchange economy, satisfying the same assumptions as in proposition 10.1.3. Let $z : T \to \mathbf{R}^{l \cdot \#N}$ be attainable net trade plans ($\{z^j + e^j\}_{j \in N} \in F^N$) such that $Eu^j(z^j + e^j \mid t^j) \geq Eu^j(e^j \mid t^j)$ for every $j \in N$ and $t^j \in T^j$. Then*

10: Existence

there exist attainable and Bayesian incentive-compatible net trade plans $\hat{z}$: $T \to \mathbf{R}^{l \cdot \#N}$ such that z and $\hat{z}$ give rise to the same interim expected utility allocation,

$$\forall j \in N : \forall t^j \in T^j : Eu^j(z^j + e^j \mid t^j) = Eu^j(\hat{z}^j + e^j \mid t^j).$$

Proof Let z be the net trade plans given in the lemma. Define $\hat{z} : T \to \mathbf{R}^{l \cdot \#N}$ by

$$\hat{z}^j(t) := \begin{cases} z^j(t) & \text{if } t \in T^*, \\ 0 & \text{otherwise.} \end{cases}$$

Clearly, $\hat{z}$ is attainable, and gives rise to the same interim expected utility allocation as z. We only need to check Bayesian incentive compatibility of $\hat{z}$. Choose any $t^j \in T^j$ and any $s^j \in T^j \setminus \{t^j\}$.

$$Eu^j(\hat{z}^j(s^j, \cdot) + e^j(t^j) \mid t^j)$$
$$= \sum_{\tau^{N \setminus \{j\}} \in T^{N \setminus \{j\}}} \pi^j(\tau^{N \setminus \{j\}} \mid t^j) u^j(\hat{z}^j(s^j, \tau^{N \setminus \{j\}}) + e^j(t^j))$$
$$= \sum_{\tau^{N \setminus \{j\}} \in T^{N \setminus \{j\}} : (t^j, \tau^{N \setminus \{j\}}) \in T^*} \pi^j(\tau^{N \setminus \{j\}} \mid t^j) u^j(\hat{z}^j(s^j, \tau^{N \setminus \{j\}}) + e^j(t^j))$$
$$= \sum_{\tau^{N \setminus \{j\}} \in T^{N \setminus \{j\}} : (t^j, \tau^{N \setminus \{j\}}) \in T^*} \pi^j(\tau^{N \setminus \{j\}} \mid t^j) u^j(e^j(t^j))$$
$$\leq Eu^j(z^j + e^j \mid t^j)$$
$$= Eu^j(\hat{z}^j + e^j \mid t^j).$$

□

Proof of Proposition 10.1.3 Let z be a coarse core allocation, whose existence is asserted by Wilson (the same assertion as proposition 10.1.1, except that private measurability and Bayesian incentive compatibility are not imposed). In the light of lemma 10.1.4, there exists an attainable, Bayesian incentive-compatible allocation $\hat{z}$ which gives rise to the same *interim* expected utility allocation as z. No coalition can improve upon z using its attainable allocation, Bayesian incentive-compatible or not. So no coalition can improve upon $\hat{z}$ using its Bayesian incentive-compatible, attainable allocation. □

Non-exclusive information implies that many type-profiles are of probability 0 according to $\pi^j(\cdot \mid t^j)$; it is illustrated in figure 10.1, the Vohra box diagram. Here, $N = \{1, 2, 3\}$, and each type space has two elements, $T^j = \{t^j_1, t^j_2\}$, $j \in N$. The support of each *ex ante* probability is given as

$$T^* = \{(t^1_1, t^2_1, t^3_2), \ (t^1_2, t^2_1, t^3_1), \ (t^1_1, t^2_2, t^3_2)\},$$

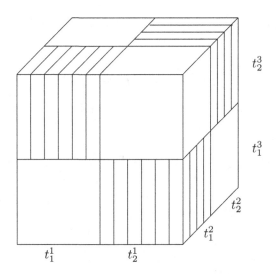

Figure 10.1: The Vohra box diagram

indicated by the shaded area in figure 10.1. Information is obtained not only through the private information structure but also through the *interim* probability.

We turn to the works on the *interim* Bayesian incentive-compatible core. It is easy to establish that *when $l = 1$, the zero net-trade plan $t \mapsto \mathbf{0}$ (resulting in the initial endowment function $t \mapsto \{e^j(t^j)\}_{j \in N}$ as the final allocation plan) is the unique interim Bayesian incentive-compatible core plan of a Bayesian pure exchange economy.*

For $l \geq 2$, the *interim* Bayesian incentive-compatible core may be empty, even for a Bayesian pure exchange economy satisfying all the neoclassical convexity assumptions. See, e.g., Hahn and Yannelis (1997, corollary 8.2.3) for an example of an economy with an empty *interim* Bayesian incentive-compatible core. The negative result is true even for some economies with linear von Neumann-Morgenstern utility functions. This negative result makes a contrast to nonemptiness of the *ex ante* Bayesian incentive-compatible core (see the proof of theorem 8.2.1 for the role of linearity of the utility functions in dealing with Bayesian incentive compatibility). Vohra (1999, Example 2.1, pp. 131-132) showed this negative result in his example of a market for a single indivisible lemon. The following example is a minor variant of Vohra's example, obtained by removing the indivisibility.

10: Existence

EXAMPLE 10.1.5 Consider the following two-consumer, two-commodity Bayesian pure exchange economy.

$$N = \{1,2\}, \quad T^1 = \{l,h\}, \quad T^2 = \{t^2\}.$$

The type-profile space $T := T^1 \times T^2$ is then identified with T^1. Consumer 1 is the seller of divisible commodity 1, and consumer 2 is the buyer. The seller knows the quality of commodity 1, but the buyer does not. The second commodity is money. The consumption set for each consumer is $\mathbf{R}_+^2$. The initial endowment function on T is a constant function,

$$\forall\, t \in T: \quad e^1(t) = \begin{pmatrix} 1 \\ 0 \end{pmatrix}, \quad e^2(t) = \begin{pmatrix} 0 \\ w \end{pmatrix},$$

where $w \geq 6$. We formulate the quality of commodity 1 in terms of the state-dependent von Neumann-Morgenstern utility function. In state l, commodity 1 is of low quality, giving rise to no utility. In state h, commodity 1 is of high quality, giving rise to positive utility. The utility functions are given by:

$$u^1(c,t) := \begin{cases} c_2 & \text{if } t = l \\ 10c_1 + c_2 & \text{if } t = h, \end{cases}$$

$$u^2(c,t) := \begin{cases} c_2 & \text{if } t = l \\ 15c_1 + c_2 & \text{if } t = h. \end{cases}$$

The *ex ante* probability held by the uninformed buyer is the uniform probability,

$$\pi^2(l,t^2) = \pi^2(h,t^2) = \frac{1}{2}.$$

We will show that this economy has no *interim* Bayesian incentive-compatible core net-trade plan. (The readers may skip the rest of this example; we will briefly comment on emptiness of the core of this economy in remark 10.1.6.)

By abuse of notation, when we have a constant plan z^j, we also use the notation z^j for the image of the function. The same abuse of notation for e^j. Let z^1 be a constant function such that $z^1 \geq -e^1$. Then,

$$u^1(z^1 + e^1, t) = \begin{cases} z_2^1 & \text{if } t = l \\ 10(1 + z_1^1) + z_2^1 & \text{if } t = h. \end{cases}$$

For a constant function z^2 for which $z^2 \geq -e^2$,

$$Eu^2(z^2 + e^2) = 7.5z_1^2 + z_2^2 + w.$$

The proof of Hahn and Yannelis' proposition (proposition 4.1.3) actually establishes that the private measurable net trade plans for which the total demand is met by the total supply with strict equality at any type profile are constant functions. This fact and Ichiishi and Radner's lemma (lemma 9.2.10) guarantee that we can consider only constant plans.

In the grand coalition N, consumer 1's (constant) strategy has the range

$$\begin{pmatrix} -1 \\ 0 \end{pmatrix} \leq z^1 \leq \begin{pmatrix} 0 \\ w \end{pmatrix},$$

and consumer 2's strategy is given by $z^2 = -z^1$.

We consider the individual rationality, viz., the conditions on a constant strategy bundle $z \in F'^N$ for no singleton coalition to *interim* block. Coalition $\{1\}$ cannot block at $t = l$, iff

$$z_2^1 \geq 0.$$

Coalition $\{1\}$ cannot block at $t = h$, iff

$$10(1 + z_1^1) + z_2^1 \geq 10.$$

Coalition $\{2\}$ cannot block, iff

$$-7.5 z_1^1 - z_2^1 + w \geq w.$$

The only z^1 which satisfies the above three inequalities is the no-trade: $z^{*1} = \mathbf{0}$. This is, therefore, the only *possible interim* core strategy.

We will show that the no-trade z^{*1} is blocked by N at $t = l$, which will complete the proof that this specific model has an empty core. (The no-trade cannot be blocked by N at $t = h$, but this fact is immaterial.) Let $z \in F'^N$ be a constant strategy proposed in order to block z^{*1}. Consumer 2 agrees to the blocking, iff

$$-7.5 z_1^1 - z_2^1 + w > w.$$

Consumer 1 agrees to the blocking at $t = l$, iff

$$z_2^1 > 0.$$

Both agree to the blocking at $t = l$ iff these two inequalities are satisfied. There are many z^1 satisfying these two inequalities, e.g.,

$$z^1 = \begin{pmatrix} -1 \\ 6 \end{pmatrix}.$$

□

REMARK 10.1.6 The key idea in example 10.1.5 may succinctly be captured by the following observation. The no-trade *interim* utility allocation is

$$u^1(e^1, t) = \begin{cases} 0 & \text{if } t = l \\ 10 & \text{if } t = h \end{cases}$$
$$Eu^2(e^2) = w.$$

The constant strategy bundle $z \in \hat{F}^N$ given by

$$z^1(l) = z^1(h) := \begin{pmatrix} -1 \\ 6 \end{pmatrix},$$
$$z^2(t^2) := -z^1(l)$$

gives rise to the *interim* utility allocation,

$$u^1(z^1 + e^1, t) = \begin{cases} 6 & \text{if } t = l \\ 6 & \text{if } t = h \end{cases}$$
$$Eu^2(z^2 + e^2) = 1.5 + w.$$

The no-trade utility allocation is blocked by N at $t = l$ via z, and the utility allocation of z is blocked by $\{1\}$ at $t = h$ via the no-trade. □

Remark 10.1.6 suggests the fact that an *interim* Bayesian incentive-compatible core strategy bundle may not exist, because a utility allocation is blocked by a coalition at a particular type profile (S, t^S), and another utility allocation which is stable against (S, t^S) is blocked by a coalition at another type profile $(S', t'^{S'})$. In order to guarantee the existence of a core strategy bundle, therefore, we need to explore the effects of different type profiles. The rest of this section will explore conditions among effects of different type profiles under which a core strategy bundle does exist.

For a clear-cut result, we will postulate that $T(\pi^j) = T$ for all j, and that von Neumann-Morgenstern utility functions are affine linear (assumption 10.1.7). In the following, commodity bundles $c^j + e^j(t^j)$ are understood as l-dimensional column vectors.

ASSUMPTION 10.1.7 (Risk Neutrality) For each consumer j and each type t^j, there exist a nonnegative row vector $a^j(t^j)$ and a scaler $b^j(t^j)$ such that

$$Eu^j\left(c^j + e^j \mid t^j\right) = a^j(t^j)\left(c^j + e^j(t^j)\right) + b^j(t^j), \quad \text{for all} \quad c^j \in \mathbf{R}^l.$$

Here, $a^j(t^j)$ is the vector of marginal (*interim*) utilities given type t^j, postulated to be nonnegative.

Define the attainable choice space C_0^S as:

$$C_0^S := \left\{ c^S \in \mathbf{R}^{l \cdot \#S} \;\middle|\; \begin{array}{l} \forall\, j \in S : \forall\, t^j \in T^j : c^j + e^j(t^j) \geq \mathbf{0} \\ \sum_{j \in S} c^j \leq \mathbf{0} \end{array} \right\}$$

$$= \left\{ c^S \in \mathbf{R}^{l \cdot \#S} \;\middle|\; \begin{array}{l} \forall\, j \in S : c^j + \underline{e}^j \geq \mathbf{0} \\ \sum_{j \in S} c^j \leq \mathbf{0} \end{array} \right\},$$

where $\underline{e}_h^j := \min\{e_h^j(t^j) \mid t^j \in T^j\}$, in short, $\underline{e}^j := \inf_{t^j \in T^j} e^j(t^j)$. The set C_0^S is nonempty; indeed, $\mathbf{0} \in C_0^S$.

An *agent* of economy $\mathcal{E}_{pe}$ is defined as a consumer together with his type, (j, t^j); denote by $\mathbf{A}$ the set of all agents,

$$\mathbf{A} := \{(j, t^j) \mid j \in N, t^j \in T^j\}.$$

An *admissible blocking coalition* is a coalition of agents in which at most one agent represents each consumer; denote by $\mathcal{B}_0$ the family of all admissible blocking coalitions,

$$\mathcal{B}_0 := \{B \subset \mathbf{A} \mid [(i, t^i), (j, t^j) \in B, t^i \neq t^j] \Rightarrow i \neq j\}.$$

Thus, consumer-coalition S forms as a blocking coalition in $\mathcal{E}_{pe}$ at type profile $\bar{t}^S$, iff the admissible agent-coalition $B := \{(j, \bar{t}^j) \in \mathbf{A} \mid j \in S\}$ forms. For $B \in \mathcal{B}_0$, let $S(B)$ be the set of those consumers represented by the agents B,

$$S(B) := \{j \in N \mid \exists\, t^j \in T^j : (j, t^j) \in B\}.$$

Also, let $t^j(B)$ be the consumer j's type for which $(j, t^j(B)) \in B$.

In the light of the linearity assumption (assumption 10.1.7), we may define the *maximal coalitional gain* for each $B \in \mathcal{B}_0$,

$$v(B) := \max \left\{ \sum_{j \in S(B)} a^j(t^j(B)) c^j \;\middle|\; \begin{array}{l} c^S \in C_0^{S(B)}, \\ \forall\, j \in S(B) : a^j(t^j(B)) c^j \geq 0 \end{array} \right\}.$$

It is achieved with net trades within $S(B)$ that are individually feasible $(c^j + \underline{e}^j \geq \mathbf{0})$, coalitionally attainable $(\sum_{j \in S(B)} c^j \leq \mathbf{0})$, and individually rational $(a^j(t^j(B)) c^j \geq 0)$. The concept of maximal coalitional gain $v(B)$ assumes transfer of utilities among the players, but these numerical values are needed only in the following quantitative condition (assumption 10.1.8) of the existence theorem (theorem 10.1.9). Notice that the gain $v(B)$ depends upon $\{\underline{e}^j\}_{j \in S(B)}$. We will discuss assumption 10.1.8 after presentation of theorem 10.1.9.

ASSUMPTION 10.1.8 (Possibility of Utility Enhancing Multilateral Trades) For all $\{\lambda_B\}_{B\in\mathcal{B}_0}$ $(\subset \mathbf{R}_+)$ and all $\{\mu^j\}_{j\in N}$ $(\subset \mathbf{R}_+^l)$ for which

$$\forall\, i, j \in N :$$
$$\sum_{B\in\mathcal{B}_0:S(B)\ni i} \lambda_B a^i(t^i(B)) + \mu^i = \sum_{B\in\mathcal{B}_0:S(B)\ni j} \lambda_B a^j(t^j(B)) + \mu^j, \quad (10.2)$$

it follows that

$$\sum_{B\in\mathcal{B}_0} \lambda_B v(B) \leq \sum_{j\in N} \mu^j \underline{e}^j. \quad (10.3)$$

THEOREM 10.1.9 (Ichiishi and Yamazaki, 2004) *Let $\mathcal{E}_{pe}$ be a Bayesian pure exchange economy such that $T(\pi^j) = T$ for all j, in which each consumer's strategy is a net-trade plan. Assume $\mathcal{E}_{pe}$ satisfies assumptions 10.1.7 and 10.1.8. Then an interim Bayesian incentive-compatible core net-trade plan of $\mathcal{E}_{pe}$ exists.*

To clarify the meaning of assumption 10.1.8, consider for example the economy $\mathcal{E}_{pe}$ with two consumers ($N = \{1, 2\}$). If $v(B) = 0$ for all $B \in \mathcal{B}_0$, then the assumption is automatically satisfied, so the *interim* Bayesian incentive-compatible core is nonempty. Otherwise, for each consumer j, let K^j be the cone spanned by consumer j's marginal (*interim*) utility vectors,

$$K^j := \left\{ \sum_{B\in\mathcal{B}_0:S(B)\ni j} \lambda_B a^j(t^j(B)) \in \mathbf{R}_+^l \;\middle|\; (\forall\, B \in \mathcal{B}_0 : S(B) \ni j) : \lambda_B \geq 0 \right\}.$$

If there exists nonzero $\{\lambda_B\}_{B\in\mathcal{B}_0}$ which gives rise to a member in $K^1 \cap K^2$, then together with $\mu^j = \mathbf{0}$ for all $j \in N$, it satisfies constraint (10.2). So the required inequality (10.3) is not satisfied unless $v(B) = 0$ for all $B \in \mathcal{B}_0$, and theorem 10.1.9 cannot be applied. Figure 10.2 illustrates this point using example 10.1.5 of a market for divisible lemons, which partially explains nonexistence of the *interim* Bayesian incentive-compatible net-trade plan in this example.

If $K^1 \cap K^2 = \{\mathbf{0}\}$, then for any nontrivial λ_B's and μ^j's to satisfy (10.2), some μ^j must be nonzero, and if the corresponding $\underline{e}^j$ is large, the required inequality (10.3) is satisfied, and the Bayesian incentive-compatible *interim* core is nonempty.

Here is *one* economic interpretation of assumption 10.1.8. Define

$$\nu := \sum_{B\in\mathcal{B}_0:S(B)\ni j} \lambda_B a^j(t^j(B)) + \mu^j \in \mathbf{R}_+^l.$$

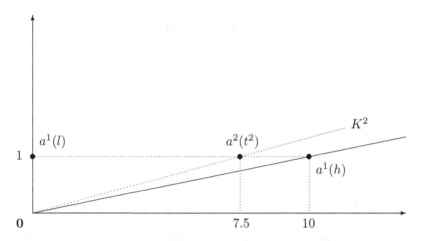

Figure 10.2: Market for divisible lemons

The vector ν is independent of j in view of (10.2). Suppose that the society values each agent-coalition B as λ_B, and each commodity h as ν_h. Then, by holding a unit of the hth initial endowment, consumer j enjoys two attributes: One is the increase in utility by possessing it as a member of various coalitions, $\sum_{B \in \mathcal{B}_0 : S(B) \ni j} \lambda_B a_h^j(t^j(B))$, and the other is its excess value as an asset, $\mu_h^j := \nu_h - \sum_{B \in \mathcal{B}_0 : S(B) \ni j} \lambda_B a_h^j(t^j(B))$. Assumption 10.1.8 says that in this situation, the society's maximal utility gain, re-scaled in order to take into account the value of each coalition, $\sum_{B \in \mathcal{B}_0} \lambda_B v(B)$, is achieved by the total excess value of the initial endowments, $\sum_{j \in N} \mu^j \underline{e}^j$.

Proof of Theorem 10.1.9 As in example 10.1.5, we may consider only constant net trade plans. A constant strategy bundle $t \mapsto c^*$ is feasible in the grand coalition, if

$$\forall j \in N: \quad c^{*j} \geq -\underline{e}^j, \tag{10.4}$$

$$-\sum_{i \in N} c^{*i} \geq \mathbf{0}. \tag{10.5}$$

(Recall $\underline{e}^j := \inf_{t^j \in T^j} e^j(t^j)$.) It is coalitionally stable, if

$$\forall B \in \mathcal{B}_0: \quad \sum_{j \in S(B)} a^j(t^j(B)) c^{*j} \geq v(B). \tag{10.6}$$

A constant map $T \to (\mathbf{R}^l)^N, t \mapsto c^*$ is the required *interim* Bayesian incentive-compatible core net-trade plan, if c^* satisfies the above linear inequality system (10.4)–(10.6). By a version[1] of the Minkowski-Farkas lemma, we can obtain a necessary and sufficient condition for the existence of such c^*. In the following, $\lambda := \{\lambda_B\}_{B \in \mathcal{B}_0}$, $\mu := \{\mu^j\}_{j \in N}$, and ν are $\#\mathcal{B}_0$-dimensional, $(\#N)l$-dimensional, and l-dimensional row vectors, respectively (each μ^j is a l-dimensional subvector). The linear inequality system (10.4)–(10.6) has a solution, iff

$$\left(\forall\, (\lambda, \mu, \nu) \in \mathbf{R}_+^{(\#\mathcal{B}_0)+(\#N)l+l} : \right.$$

$$\sum_{B \in \mathcal{B}_0 : S(B) \ni j} \lambda_B a^j(t^j(B)) + \mu^j - \nu = \mathbf{0} \text{ for all } j \in N \Bigg) :$$

$$\sum_{B \in \mathcal{B}_0} \lambda_B v(B) - \sum_{j \in N} \mu^j \underline{e}^j - \nu \mathbf{0} \leq 0.$$

By eliminating ν, and by observing that $a^j(t^j(B))$, λ_B and μ^j are nonnegative, we obtain the condition of theorem 10.1.9. □

The rest of this section is devoted to analysis of several variants of the example of a market for divisible lemons (example 10.1.5).

EXAMPLE 10.1.10 We specify N, T, and u^1 as in example 10.1.5, but consider different data on consumer 2's expected utility $Eu^2 : \mathbf{R}_+^2 \to \mathbf{R}$ (still assumed to be affine linear, so that $Eu^2(c) = a^2(t^2)c$) and on the constant initial endowment vectors $e^j := (e_1^j, e_2^j) \in \mathbf{R}_+^2$, $j \in N$.

Define for simplicity,

$$(N, l) := \{(1, l), (2, t^2)\} \in \mathcal{B}_0, \text{ and}$$
$$(N, h) := \{(1, h), (2, t^2)\} \in \mathcal{B}_0.$$

In considering λ_B's and μ^j's satisfying condition (10.2), we may assume $\lambda_B = 0$ for all B for which $\#S(B) = 1$, and may set $\lambda := \lambda_{(N,l)}$, and $\lambda_{(N,h)} = 1 - \lambda$ in the light of homogeneity.

Case 1. Suppose $a^2(t^2) = (\alpha, 1)$, for some $\alpha > 10$. See figure 10.3. Then, $K^1 \cap K^2 = \{\mathbf{0}\}$. If $10e_1^1 \leq e_2^1$, then

$$v(B) = \begin{cases} \alpha e_1^1 & \text{for } B = (N, l), \\ (\alpha - 10)e_1^1 & \text{for } B = (N, h) \end{cases}$$

(both maximal gains $v(B)$, $B = (N, l), (N, h)$, can be achieved by the individually feasible, coalitionally attainable and individually rational net

---
[1] We use the following version: Let A be an $m \times n$ matrix, and let b be a $m \times 1$ matrix. Then, there exists $x \in \mathbf{R}^n$ such that $Ax \geq b$, iff for every $1 \times m$ matrix $\lambda \geq \mathbf{0}$ for which $\lambda A = \mathbf{0}$ it follows that $\lambda b \leq 0$.

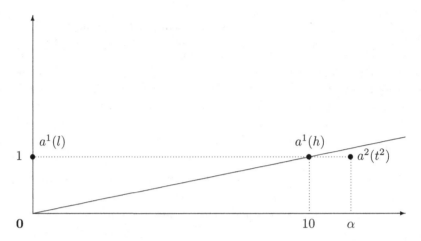

Figure 10.3: Example 10.1.10, Case 1. $a^2(t^2) = (\alpha, 1)$, $\alpha > 10$

trades $c^1 = (-e_1^1, 10e_1^1)$, $c^2 = -c^1$). Coefficients λ and μ^j's satisfy (10.2), iff

$$(1-\lambda)10 + \mu_1^1 = \alpha + \mu_1^2$$
$$1 + \mu_2^1 = 1 + \mu_2^2.$$

The required inequality (10.3) becomes

$$\lambda \alpha e_1^1 + (1-\lambda)(\alpha - 10)e_1^1 \leq \mu^1 e^1 + \mu^2 e^2,$$

that is,

$$[\alpha - (1-\lambda)10 - \mu_1^1]e_1^1 \leq \mu_2^1 e_2^1 + \mu^2 e^2.$$

But by (10.2), the left-hand side is equal to $-\mu_1^2 e_1^1$, so the required inequality is always satisfied.

If $10e_1^1 > e_2^2$, then

$$v(B) = \begin{cases} \alpha e_1^1 & \text{for } B = (N, l), \\ (\alpha - 10)\frac{1}{10}e_2^2 < (\alpha - 10)e_1^1 & \text{for } B = (N, h), \end{cases}$$

so given (10.2), the required condition is *a fortiori* satisfied.

Thus, assumption 10.1.8 is satisfied for arbitrary initial endowment vectors, e^1, e^2.

Case 2. Suppose $a^2(t^2) = (15, 0)$. Then, $K^1 \cap K^2 = \{0\}$. If $10e_1^1 \leq e_2^2$,

$$v(B) = \begin{cases} 15e_1^1 + e_2^2, & \text{for } B = (N, l), \\ 5e_1^1 + e_2^2, & \text{for } B = (N, h) \end{cases}$$

(both maximal gains can be achieved by the individually feasible, coalitionally attainable and individually rational net trades $c^1 = (-e_1^1, e_2^2)$, $c^2 = -c^1$). Coefficients λ and μ^j's satisfy (10.2), iff

$$(1-\lambda)10 + \mu_1^1 = 15 + \mu_1^2$$
$$1 + \mu_2^1 = \mu_2^2.$$

The required inequality (10.3) becomes

$$\lambda(15e_1^1 + e_2^2) + (1-\lambda)(5e_1^1 + e_2^2) \leq \mu^1 e^1 + \mu^2 e^2,$$

that is,

$$[15 - (1-\lambda)10 - \mu_1^1]e_1^1 + e_2^2 \leq \mu_2^1 e_2^1 + \mu^2 e^2.$$

In view of (10.2), this is

$$-\mu_1^2 e_1^1 + e_2^2 \leq \mu_2^1 e_2^1 + \mu_1^2 e_1^2 + (1+\mu_2^1)e_2^2,$$

which is always satisfied.

If $10e_1^1 > e_2^2$,

$$v(B) = \begin{cases} 15e_1^1 + e_2^2, & \text{for } B = (N, l), \\ 5 \cdot \frac{1}{10}e_2^2 + e_2^2 < 5e_1^1 + e_2^2, & \text{for } B = (N, h), \end{cases}$$

so given (10.2), the required condition is *a fortiori* satisfied.

Thus, assumption 10.1.8 is satisfied for arbitrary initial endowment vectors, e^1, e^2.

Case 3. Suppose $a^2(t^2) = (7.5, 0)$. Then, $K^1 \cap K^2 = \{\mathbf{0}\}$, and

$$v(B) = \begin{cases} 7.5e_1^1 + e_2^2, & \text{if } B = (N, l), \\ e_2^2, & \text{if } B = (N, h) \end{cases}$$

(the second identity is due to the individual rationality of consumer 2). Coefficients λ and μ^j's satisfy (10.2), iff

$$(1-\lambda)10 + \mu_1^1 = 7.5 + \mu_1^2$$
$$1 + \mu_2^1 = \mu_2^2.$$

The required inequality (10.3) becomes

$$\lambda\left(7.5e_1^1 + e_2^2\right) + (1-\lambda)e_2^2 \leq \mu^1 e^1 + \mu^2 e^2.$$

On the one hand, condition (10.3) is satisfied, if $e_1^1 = 0$. On the other hand, if $e_1^1 > 0$, condition (10.3) is violated by $\lambda = 0.25$, $\mu^1 = \mathbf{0}$ and $\mu^2 = (0, 1)$. Thus, assumption 10.1.8 is satisfied, iff $e_1^1 = 0$, that is, iff consumer 1 does not initially possess commodity 1. □

EXAMPLE 10.1.11 This is another variant of example 10.1.5. We introduce the third commodity which is also a lemon,[2] but its quality has the reverse contingency, that is, the quality is high if $t = l$, and is low if $t = h$. Thus, the two lemons (commodities 1 and 3) might kill their destabilizing effects each other. We will see, however, that assumption 10.1.8 is still violated. This means that the destabilizing effect of the very asymmetry in information is robust.

Define $a^j(t^j) \in \mathbf{R}^3$ as in assumption 10.1.7. Thus,

$$N = \{1,2\}, \quad T^1 = \{l,h\}, \quad T^2 = \{t^2\},$$

$$\forall\, t \in T: \ e^1(t) = \begin{pmatrix} 1 \\ 0 \\ 1 \end{pmatrix}, \ e^2(t) = \begin{pmatrix} 0 \\ w \\ 0 \end{pmatrix},$$

$$\begin{aligned} a^1(l) &= (0,1,10), \\ a^1(h) &= (10,1,0), \\ a^2(t^2) &= (a,1,b), \end{aligned}$$

where $0 < a < 10$ and $0 < b < 10$. We will show that this economy does not satisfy assumption 10.1.8.

Define (N,l) and (N,h) as in example 10.1.10. Also as in example 10.1.10, in considering λ_B's and μ^j's satisfying condition (10.2), we may assume $\lambda_B = 0$ for all B for which $\#S(B) = 1$, and may set $\lambda := \lambda_{(N,l)}$, and $\lambda_{(N,h)} = 1 - \lambda$. Observe that

$$v(B) = \begin{cases} a, & \text{if } B = (N,l), \\ b, & \text{if } B = (N,h) \end{cases}$$

(the maximal gain $v((N,l))$ is achieved by the net trades, $c^1 = (-1,0,0)$, $c^2 = -c^1$, which are individually feasible, coalitionally attainable and individually rational, and $v((N,h))$ is achieved by the net trades, $c^1 = (0,0,-1)$, $c^2 = -c^1$).

Condition (10.2) becomes:

$$\begin{aligned} (1-\lambda)10 + \mu_1^1 &= a + \mu_1^2, \\ 1 + \mu_2^1 &= 1 + \mu_2^2, \\ \lambda 10 + \mu_3^1 &= b + \mu_3^2. \end{aligned}$$

---

[2] In general, if there are more than two possible lemons, the type space needs to be expanded to take into account the quality of the second lemon/peach. Actually, due to the perfect (negative) correlation, we do not have to expand the dimension in the present formulation.

10: Existence

The requirement (condition (10.3)) becomes:

$$\lambda a + (1-\lambda)b \leq \mu_1^1 + \mu_3^1 + \mu_2^2 w,$$

which, in the light of (10.2), is equivalent to:

$$10 \leq (1-\lambda)a + \lambda b + \mu_1^2 + \mu_3^2 + \mu_2^2 w.$$

We will first show that $10 < a+b$, if (10.3) is to be valid for all (λ, μ^1, μ^2) satisfying (10.2). Indeed, if $10 \geq a + b$, then there exists λ satisfying

$$\begin{aligned}(1-\lambda)10 &\geq a \\ \lambda 10 &\geq b.\end{aligned}$$

Set $\mu_1^1 = \mu_3^1 = \mu_2^i = 0$, and $\mu_1^2 := (1-\lambda)10 - a \geq 0$, $\mu_3^2 := \lambda 10 - b \geq 0$. These λ, μ^1, μ^2 satisfy condition (10.2). But (10.3) becomes

$$\lambda a + (1-\lambda)b \leq 0,$$

which cannot be true.

It suffices to show also that $10 > a + b$, if (10.3) is to be valid for all (λ, μ^1, μ^2) satisfying (10.2). Indeed, if $10 \leq a + b$, then there exists λ satisfying

$$\begin{aligned}(1-\lambda)10 &\leq a \\ \lambda 10 &\leq b,\end{aligned}$$

Set $\mu_1^2 = \mu_3^2 = \mu_2^i = 0$, and $\mu_1^1 := a - (1-\lambda)10 \geq 0$, $\mu_3^1 := b - \lambda 10 \geq 0$. These λ, μ^1, μ^2 satisfy condition (10.2). But (10.3) becomes

$$10 \leq \lambda a + (1-\lambda)b,$$

which cannot be true. □

In addition to the negative result on the single lemon market example, we have obtained a partially negative result (case 3 of example 10.1.10) and a totally negative result (example 10.1.11). We turn to positive examples 10.1.12 – 10.1.14 now. We will show in example 10.1.12 that a slight modification of the previous examples (case 3 of example 10.1.10, and example 10.1.11) guarantees assumption 10.1.8. Examples 10.1.13 and 10.1.14 are modified cases 1 and 2 of example 10.1.10, and are included here for completeness.

EXAMPLE 10.1.12 We turn to case 3 of example 10.1.10 and to example 10.1.11, both of which either had no endowment of the first commodity (lemon) by a potential seller (i.e., $e_1^1 = 0$) or did not satisfy assumption 10.1.8. It is worthwhile to see what kind of perturbation is needed to change those examples into ones that satisfy the assumption of the theorem while requiring $e_1^1 > 0$. The common element in both of these examples is that the uninformed consumer does not give absolutely higher evaluation of a potential lemon than the informed consumer in high quality state in terms of their conditional expected utilities. Thus, it would be very interesting to check whether our assumption can still be satisfied under these circumstances in a variant of the single lemon model.

In case 3 of example 10.1.10 the second commodity is no longer a "money" but a commodity that gives possibly different degrees of marginal utilities to consumers. It seems immediately clear to us that one of the key factors in this example is that consumer 2 who is a potential "buyer" in the lemon's model must have at least relatively higher conditional marginal utility for a potential lemon than consumer 1 who is a potential "seller." We simply modify the example in case 3 as follows: The initial endowment function on T is a constant function,

$$\forall t \in T: \ e^1(t) = \begin{pmatrix} e_1^1 \\ e_2^1 \end{pmatrix}, \ e^2(t) = \begin{pmatrix} e_1^2 \\ e_2^2 \end{pmatrix} \text{ with } e_1^1 > 0 \text{ and } e_2^2 > 0.$$

Let [3]

$$\begin{aligned} a^1(l) &= (0, \beta_l), \\ a^1(h) &= (\alpha_h, \beta_h), \\ a^2(t^2) &= (\alpha, \beta), \end{aligned}$$

and

$$0 < \alpha < \alpha_h,$$
$$0 < \beta < \min\{\beta_l, \beta_h\},$$
$$MRS_{12}^2 = \tfrac{\alpha}{\beta} > \tfrac{\alpha_h}{\beta_h} = MRS_{12}^{1h}.$$

The remaining specification and notation are as in example 10.1.10. Then, $K^1 \cap K^2 = \{0\}$, and we have

$$\begin{aligned} v((N,l)) &= \beta_l\left(\frac{\alpha}{\beta}\right) \min\left\{e_1^1, \frac{\beta}{\alpha}e_2^2\right\} \\ v((N,h)) &= \beta_h\left(\frac{\alpha}{\beta} - \frac{\alpha_h}{\beta_h}\right) \min\left\{e_1^1, \frac{\beta}{\alpha}e_2^2\right\}. \end{aligned}$$

---

[3] MRS_{12}^j denotes the consumer j's marginal rate of substitution of commodity 2 for commodity 1.

Coefficients λ and μ^j's satisfy (10.2), iff
$$(1-\lambda)\alpha_h + \mu_1^1 = \alpha + \mu_1^2,$$
$$\lambda\beta_l + (1-\lambda)\beta_h + \mu_2^1 = \beta + \mu_2^2.$$

The required inequality (10.3) becomes
$$\left[\frac{\alpha}{\beta}\beta_\lambda - (1-\lambda)\alpha_h\right] \min\left\{e_1^1, \frac{\beta}{\alpha}e_2^2\right\}$$
$$\leq \left[\alpha - (1-\lambda)\alpha_h + \mu_1^2\right]e_1^1 + (\beta_\lambda - \beta + \mu_2^1)e_2^1 + Res(\mu),$$

where
$$\beta_\lambda := \lambda\beta_l + (1-\lambda)\beta_h,$$
$$Res(\mu) := \mu_1^2 e_1^1 + \mu_2^1 e_2^1 + \mu_1^2 e_1^2 + \mu_2^1 e_2^2.$$

If $\frac{\alpha}{\beta}e_1^1 > e_2^2$, condition (10.3) becomes:
$$0 \leq \left(e_1^1 - \frac{\beta}{\alpha}e_2^2\right)[\alpha - (1-\lambda)\alpha_h] + Res(\mu),$$

which need not be satisfied for values of $0 \leq \lambda < 1 - (\alpha/\alpha_h)$.

On the other hand, if $\frac{\alpha}{\beta}e_1^1 \leq e_2^2$, then condition (10.3) becomes:
$$0 \leq (\beta_\lambda - \beta)\left(e_2^2 - \frac{\alpha}{\beta}e_1^1\right) + Res(\mu),$$

which is satisfied for any $0 \leq \lambda \leq 1$ and any $\mu_i^j \geq 0$.

Therefore, assumption 10.1.8 is satisfied for parameter values of e_1^1, e_2^2 such that $\frac{\alpha}{\beta}e_1^1 \leq e_2^2$. This means that assumption 10.1.8 is satisfied when there are sufficiently many units of commodity 2 (in the hand of potential buyer of the potential lemon) so that the buyer has enough to compensate the seller for his transfer of the initial endowment of the lemon to the buyer in state (h, t^2). □

At first glance it might seem that the *interim* core is empty in case of example 10.1.12 as the expected conditional marginal utility α of the buyer for a potential lemon is strictly less than the expected conditional marginal utility α_h of the seller when it is not a lemon but a peach. Nonetheless, assumption 10.1.8 is satisfied and the *interim* core is nonempty. There are two key elements in example 10.1.12.

(a) Despite the lower expected conditional marginal utility of the buyer for a potential lemon, there is another commodity 2, his expected conditional marginal rate of substitution of which for a potential lemon is kept higher than the expected conditional marginal rate of substitution of the seller so that there is no reversal of the superiority of the expected conditional marginal rate of substitution for a potential lemon depending upon type profiles.

(b) There must be enough initial endowment of the commodity, in the hands of the buyer, that is used to compensate the seller for his transfer of his potential lemon.

Case 3 of example 10.1.10 does not satisfy (a). Example 10.1.11 does satisfy (a) but not (b).

EXAMPLE 10.1.13 We modify case 1 of example 10.1.10. Intuitively, what is "wrong" about example 10.1.5 is that the uninformed consumer does not posses a commodity which the informed consumer badly wishes to have so that there is a very profitable trade between the two. Thus, we introduce a third commodity which "enhances" the utility of the informed consumer without affecting that of the uninformed consumer. The consumption set for each consumer is $\mathbf{R}_+^3$. The initial endowment function on T is a constant function,

$$\forall\, t \in T:\ e^1(t) = \begin{pmatrix} 1 \\ 0 \\ e_3^1 \end{pmatrix},\ e^2(t) = \begin{pmatrix} 0 \\ w \\ e_3^2 \end{pmatrix}.$$

Let

$$\begin{aligned} a^1(l) &= (0, 1, \beta_l), \\ a^1(h) &= (10, 1, \beta_h), \\ a^2(t^2) &= (\alpha, 1, 0), \end{aligned}$$

and assume

$$\beta_h e_3^2 \geq 10.$$

The remaining specification and notation are as in example 10.1.10. Then, $K^1 \cap K^2 = \{\mathbf{0}\}$, and

$$\begin{aligned} v((N, l)) &= \beta_l e_3^2 + \alpha, \\ v((N, h)) &= -10 + \beta_h e_3^2 + \alpha\,. \end{aligned}$$

Coefficients λ and μ^j's satisfy (10.2), iff

$$\lambda(0, 1, \beta_l) + (1 - \lambda)(10, 1, \beta_h) + \mu^1 = (\alpha, 1, 0) + \mu^2.$$

The required inequality (10.3) becomes
$$0 \leq \mu_3^1(e_3^1 + e_3^2) + \mu_2^2 w.$$
Thus, assumption 10.1.8 is always satisfied. Note that $e_1^1 = 1$ so that the informed consumer has a single potential lemon as in example 10.1.5, and still assumption 10.1.8 is satisfied. The specific circumstance in this example that contributes to fulfillment of assumption 10.1.8 is that there is a third commodity which highly enhances the utility of the informed consumer if it is provided to him in exchange for the single potential lemon whose consumption in turn by the uninformed consumer highly enhances his utility. □

EXAMPLE 10.1.14 We turn to case 2 of example 10.1.10. Noting that in example 10.1.13 the second commodity "money" does not play any role in guaranteeing the fulfillment of the condition, a crucial aspect is to have a commodity like the third one in the example. Thus, we modify case 2 so as to have a more general setting than the previous one with the second commodity eliminated. Thus, the consumption set for each consumer is $\mathbf{R}_+^2$. The initial endowment function on T is a constant function,

$$\forall t \in T : \ e^1(t) = \begin{pmatrix} e_1^1 \\ e_2^1 \end{pmatrix}, \ e^2(t) = \begin{pmatrix} e_1^2 \\ e_2^2 \end{pmatrix}, \ \text{with } e_1^1 > 0, \ e_2^2 > 0.$$

Moreover, we assume
$$0 < \frac{\alpha}{\beta} e_1^1 \leq e_2^2.$$

Let
$$\begin{aligned} a^1(l) &= (\alpha_l, \beta_l), \\ a^1(h) &= (\alpha_h, \beta_h), \\ a^2(t^2) &= (\alpha, \beta), \end{aligned}$$

and
$$0 < \max\{\alpha_l, \alpha_h\} < \alpha,$$
$$0 < \beta < \min\{\beta_l, \beta_h\},$$
$$MRS_{12}^{1l} = \tfrac{\alpha_l}{\beta_l} < MRS_{12}^{1h} = \tfrac{\alpha_h}{\beta_h} < MRS_{12}^2 = \tfrac{\alpha}{\beta}.$$

The remaining specification and notation are as in example 10.1.10. Then, $K^1 \cap K^2 = \{0\}$, and we have

$$v((N,l)) = \beta_l \left(\frac{\alpha}{\beta} - \frac{\alpha_l}{\beta_l} \right) \min\left\{ e_1^1, \frac{\beta}{\alpha} e_2^2 \right\}$$

$$v((N,h)) = \beta_h \left(\frac{\alpha}{\beta} - \frac{\alpha_h}{\beta_h} \right) \min\left\{ e_1^1, \frac{\beta}{\alpha} e_2^2 \right\}.$$

Coefficients λ and μ^j's satisfy (10.2), iff

$$\lambda(\alpha_l, \beta_l) + (1-\lambda)(\alpha_h, \beta_h) + \mu^1 = (\alpha, \beta) + \mu^2.$$

The required inequality (10.3) becomes

$$0 \leq (\beta_\lambda - \beta)\left(e_2^2 - \frac{\alpha}{\beta}e_1^1\right) + Res(\mu),$$

where

$$\beta_\lambda := \lambda\beta_l + (1-\lambda)\beta_h,$$
$$Res(\mu) := \mu_1^2 e_1^1 + \mu_2^1 e_2^1 + \mu_1^2 e_1^2 + \mu_2^1 e_2^2.$$

The inequality is always satisfied. Therefore, assumption 10.1.8 is satisfied for parameter values of e_1^1, e_2^2 such that $\frac{\alpha}{\beta}e_1^1 \leq e_2^2$. The specific circumstance in this example that contributes to fulfillment of the condition is essentially the same as in the previous example. □

10.2 *Ex Ante* Solutions

We turn to the Bayesian incentive-compatible *ex ante* core concepts. Each consumer j in Bayesian pure exchange economy $\mathcal{E}_{pe}$ is endowed with an *ex ante* probability π^j on the type-profile space T.

The first consequence of Ichiishi and Idzik's (1996) general existence result (theorem 8.2.1) is an obvious special case: If each utility function $u^j(\cdot, t)$ of economy $\mathcal{E}_{pe}$ is affine linear on C^j, then there exists an *ex ante* Bayesian incentive-compatible core allocation in the private information case, regardless whether a choice is a commodity bundle or is a net trade.

The second consequence, which is also straightforward, is on economy $\mathcal{E}_{pe}$ in which everybody takes a mixed strategy (or rather, type-profile dependent mixed choices) and the coalitional attainability is defined as the expected feasibility. Consumer j's choice set is now the set $\mathcal{M}(C^j)$ of all probabilities on the compact consumption set C^j. His strategy is a function $\mu^j : T \to \mathcal{M}(C^j)$, $t \mapsto \mu^j[t](\cdot)$; denote by X^j the set of all such functions. His preference relation is represented by the expected utility function $U^j : \mathcal{M}(C^j) \times T \to \mathbf{R}$ defined by

$$U^j(p^j, t) := \int_{C^j} u(c^j, t) p^j(dc^j).$$

Coalition S's feasible strategy bundle is a T^S-measurable member of X^S such that for each t the expected total demand is less than or equal to the total supply. Coalition S's feasible strategy correspondence is the constant correspondence on X that takes the value,

$$F^S := \left\{ \mu^S : T \to \prod_{j \in S} \mathcal{M}(C^j) \;\middle|\; \begin{array}{l} \mu^S \text{ is } \mathcal{T}^S\text{-measurable, and for each } t, \\ \sum_{j \in S} \int_{C^j} c^j \mu^j[t](dc^j) \leq \sum_{j \in S} e^j(t) \end{array} \right\}.$$

We thus have the associated Bayesian society,

$$\mathcal{S} := \left(\{\mathcal{M}(C^j), T^j, U^j, \pi^j\}_{j \in N}, \; \{T(S), F^S\}_{S \in \mathcal{N}} \right).$$

COROLLARY 10.2.1 *Let $\mathcal{E}_{pe}$ be a Bayesian pure exchange economy, in which each consumption set C^j is a nonempty and compact subset of the commodity space $\mathbf{R}^l$ and each von Neumann-Morgenstern utility function $u^j(\cdot, t)$ is continuous in C^j. Assume that $T(S) = T(S')$ for all $S, S' \in \mathcal{N}$. Allow each consumer to take a mixed choice, and define the coalitional attainability as the expected feasibility at each type profile. Then there exists an ex ante Bayesian incentive-compatible core allocation.*

Proof We only have to verify that conditions (i)-(vi) of theorem 8.2.1 are all satisfied in the associate Bayesian society.

(i) The choice set $\mathcal{M}(C^j)$ endowed with the weak* topology is compact, convex, and metrizable.

(ii) Clearly, for any j and any t, $U^j(\cdot, t)$ is linear affine and continuous in $(\mathcal{M}(C^j), \text{weak}^*$ topology).

(iv) and (vi) Correspondence $t \mapsto F^S$, being a constant correspondence, is both upper semicontinuous and lower semicontinuous. Set F^S is clearly nonempty, closed and convex.

(v) Let $\mathcal{B}$ be a balanced family of subsets of N with associated balancing coefficients $\{\lambda_S\}_{S \in \mathcal{B}}$. Choose any $\{\mu^{S,j}\}_{j \in S} \in F^S$ and define $\nu^j := \sum_{S \in \mathcal{B}: S \ni j} \lambda_S \mu^{S,j}$ for every $j \in N$. Then,

$$\begin{aligned}
\sum_{j \in N} \int_{C^j} c^j \nu^j[t](dc^j) &= \sum_{j \in N} \sum_{S \in \mathcal{B}: S \ni j} \lambda_S \int_{C^j} c^j \mu^{S,j}[t](dc^j) \\
&= \sum_{S \in \mathcal{B}} \lambda_S \sum_{j \in S} \int_{C^j} c^j \mu^{S,j}[t](dc^j) \\
&\leq \sum_{S \in \mathcal{B}} \lambda_S \sum_{j \in S} e^j(t) \\
&= \sum_{j \in N} e^j(t).
\end{aligned}$$

So $\nu \in F^N$. $\square$

We have reviewed that affine linearity of j's utility function on C^j plays an essential role in establishing the existence of a Bayesian incentive compatible core allocation in the private information case (see the proof of theorem 8.2.1). This point is valid specifically in the Bayesian pure exchange economy when each consumer chooses a demand plan as his strategy. However, the story is different in the situation in which each consumer chooses a net trade plan as his strategy; see Forges, Minelli and Vohra (2002, proposition 1) for the following theorem.

THEOREM 10.2.2 *Let $\mathcal{E}_{pe}$ be a Bayesian pure exchange economy, in which each consumption set C^j is the nonnegative orthant $\mathbf{R}_+^l$ of the commodity space, and each von Neumann-Morgenstern utility function $u^j(\cdot, t)$ is concave, weakly monotone and continuous in C^j. Assume that $T(S) = T(S')$ for all $S, S' \in \mathcal{N}$. Suppose that each consumer chooses a net trade plan as a strategy. Then there exists an ex ante Bayesian incentive-compatible core allocation.*

Proof Define a non-side-payment game $V' : \mathcal{N} \to \mathbf{R}^N$ by

$$V'(S) := \left\{ u \in \mathbf{R}^N \;\middle|\; \begin{array}{l} \exists\, z^S : T \to \mathbf{R}^{l \cdot \#S} : \\ \forall\, j \in S : z^j \text{ is } \mathcal{T}^j\text{-measurable}, \\ \forall\, t \in T : \; z^j(t^j) + e^j(t^j) \in \mathbf{R}_+^l, \\ \quad\quad \sum_{j \in S} z^j(t^j) \leq \mathbf{0}, \\ \forall\, j \in S : u_j \leq Eu^j(z^j + e^j) \end{array} \right\}.$$

The sets $V'(S)$ are nonempty; indeed, the no-trade strategy bundle, $z^S(t) \equiv \mathbf{0}$, is always feasible, so gives rise to a member of $V'(S)$. By direct application of Scarf's theorem for nonemptiness of the core (theorem 8.A.2 of this book), the core of this game V' is nonempty. Let $z^\dagger$ be a net trade plan bundle which gives rise to a member of the core. By Ichiishi and Radner's lemma (lemma 9.2.10), there exists a net trade plan bundle z^* such that each z^{*j} is $\mathcal{T}^j$-measurable, $z^{\dagger j} \leq z^{*j}$, and $\sum_{j \in N} z^{*j}(t^j) = \mathbf{0}$, for all $t \in T$. By weak monotonicity of $u^j(\cdot, t)$, the bundle z^* gives rise to a member of the core of V'. By Hahn and Yannelis' proposition (proposition 4.1.3), each strategy z^{*j} is Bayesian incentive-compatible. □

REMARK 10.2.3 Forges and Minelli (2001) took the mediator-based approach, and constructed another kind of Bayesian pure exchange economy with probabilistic choices. Let $\mathcal{E}_{pe}$ be the deterministic Bayesian pure exchange economy (example 2.2.1). For each type profile t, let $\mathbf{C}_0^S(t)$ be the compact set of commodity bundles attainable in coalition S,

$$\mathbf{C}_0^S(t) := \left\{ c^S \in C^S \;\middle|\; \sum_{j \in S} c^j \leq \sum_{j \in S} e^j(t) \right\},$$

and denote by $\mathcal{M}(\mathbf{C}_0^S(t))$ the set of all probabilities on $\mathbf{C}_0^S(t)$. An element of $\mathcal{M}(\mathbf{C}_0^S(t))$ is not *mixed* choices, but a *correlated* choice. The product probability of mixed strategies is a correlated strategy, but in general a correlated strategy cannot be expressed as the product probability of mixed strategies. A coalitional feasible strategy is defined as a selection $\mu^S : t \mapsto \mu^S[t](\cdot)$ of the correspondence, $t \mapsto \mathcal{M}(\mathbf{C}_0^S(t))$. Given coalitional strategy μ^S, consumer j's utility at the true type profile $\bar{t}$ is

$$U^j(\mu^S[\bar{t}],\bar{t}) := \int_{C^j} u^j(c^j,\bar{t})\mu^S[\bar{t}](dc).$$

In this correlated choice framework, individuals cannot singly choose *his* probability on commodity bundles. His "action" is defined as a report of his type. If consumer j misrepresent his type as $\tilde{t}^j$, assuming that everybody else supplies honest reports, j's utility becomes

$$U^j(\mu^S[\tilde{t}^j,\bar{t}^{N\setminus\{j\}}],\bar{t}) := \int_{C^j} u^j(c^j,\bar{t})\mu^S[\tilde{t}^j,\bar{t}^{N\setminus\{j\}}](dc).$$

A coalitional feasible strategy μ^S is called Bayesian incentive-compatible, if nobody benefits from misrepresenting his own type, that is,

$$\forall \bar{t} \in T : \forall j \in S : \forall \tilde{t}^j \in T^j : U^j(\mu^S[\bar{t}],\bar{t}) \geq U^j(\mu^S[\tilde{t}^j,\bar{t}^{N\setminus\{j\}}],\bar{t}).$$

Consumer j's *ex ante* expected utility of coalitional feasible strategy μ^S is given as $EU^j(\mu^S) := \sum_t U^j(\mu^S[t],t)\pi^j(t)$, and from this we can define the *ex ante* Bayesian incentive-compatible core.

Forges and Minelli (2001) established in this framework that *if $l = \#N$, each utility function $u^j(\cdot,t)$ is additively separable in C^j, e^j is a constant function, and $e_h^j = 0$ for all $h \neq j$, then the ex ante Bayesian incentive-compatible core is nonempty.*

In the absence of Forges and Minelli's restrictive assumptions, we expect that the core may be empty. While their model does not naturally fit in our model of Bayesian society (definition 2.1.3) due to the use of correlated choices, we can nevertheless imbed it in a particular Bayesian society: Redefine coalition S's feasible strategy set as

$$F^S := \left\{ \{\mu^j\}_{j\in S} \text{ defined on } T \;\middle|\; \begin{array}{l} \forall j \in S : \mu^j[t](\cdot) \in \mathcal{M}(\mathbf{C}_0^S(t)) \\ \forall i,j \in S : \mu^i[t](\cdot) = \mu^j[t](\cdot) \end{array} \right\}.$$

When the members take a strategy bundle $\{\mu^j\}_{j\in S} \in F^S$, everybody is choosing the same strategy, and this common strategy has been called a coalitional strategy, denoted by μ^S. We view that C^S is a subset of C^N

(by setting the coordinates corresponding to $N \setminus S$ to be equal to $\mathbf{0}$). In the Bayesian society

$$(\{\mathcal{M}(C^N), T^j, U^j, \pi^j\}_{j \in N}, \ \{T(S), F^S\}_{S \in \mathcal{N}})$$

constructed this way, assumptions (i)-(iv) and (vi) are satisfied (apart from the T^j-measurability requirement, whose omission does not change the theorem as pointed out in remark 8.2.2). But assumption (v) is violated, hence the possibility of empty Bayesian incentive-compatible core. To see how (v) is violated, let $\mathcal{B}$ be a balanced family of subsets of N, and let $\{\lambda_S\}_{S \in \mathcal{B}}$ be the associated balancing coefficients. For each $S \in \mathcal{B}$, choose $\{\mu^S\}_{j \in S} \in F^S$ and define $\{\nu^j\}_{j \in N}$ by $\nu^j := \sum_{S \in \mathcal{B}: S \ni j} \lambda_S \mu^S$. Then $\nu^i \neq \nu^j$ for some $i \neq j$, so $\{\nu^j\}_{j \in N} \notin F^N$. Indeed, Forges, Mertens and Vohra (2000) further specified the model by introducing "money" and by assuming that each utility function is linear in money, and provided an example which has no Bayesian incentive-compatible core.

We point out two problems concerning the correlated-choice approach: First, like the mixed-choice approach or more generally like any probabilistic approach, it avoids the question of explaining deterministic choice. Even when a player has decided on a probabilistic choice, a time will come when he has to take a definite action. Usually in real life, he acts upon his own will (which theory needs to explain), and does not leave his action up to the outcome of throwing dice.

The second problem concerns specifically the correlated-choice approach. The approach is applicable in general only to situations in which outsiders to a coalition have no influence on the insiders, like the pure exchange economy (example 2.2.1) and the coalition production economy (example 2.2.2). It is not applicable to situations in which a coalition's feasible-strategy set depends on an outsiders' strategy bundle, or a player's utility depends on a choice bundle. To see this point, consider the simple no-externality case addressed by theorem 8.2.1, re-formulated by introducing type-profile dependent correlated choices as strategies. Suppose the grand coalition is choosing a strategy $t \mapsto \mu^N[t](\cdot)$, and coalition S is going to defect. In accordance with the spirit of the strong equilibrium, the members of S passively take the outsiders' strategies as given. However, there is no way to identify the part of the strategy μ^N that the outsiders are responsible for, so the members of S do not know which strategy of $N \setminus S$ they can passively take. One might argue that S takes the marginal probability $\text{proj}_{N \setminus S} \mu^N[t] \in \mathcal{M}(C^{N \setminus S})$ as given. Yet there is no guarantee that the product of the two marginal probabilities, $\text{proj}_{N \setminus S} \mu^N[t]$ and $\text{proj}_S \mu^N[t]$, can recover the original probability, $\mu^N[t]$. The same problem occurs when a player's utility depends fully on a choice bundle. Thus, the approach fails to address intercoalitional problems in which several coalitions influence each other: the situations commonly observed in the present-day economy with organizations.

Instead of Forges and Minelli's (2001) correlated choices, we can use randomized choices defined as functions $f^j : P \to C^j$ for some probability space $(P, \mathcal{P}, p)$. From a randomized choice bundle $f^S := \{f^j\}_{j \in S}$, we derive a correlated choice as its distribution $p \circ (f^S)^{-1}$. An individual choice is re-defined as this randomized choice. This model can address the general case with externalities (although the first problem about the very moment of definite action still remains). We expect that an *ex ante* Bayesian incentive-compatible core allocation existence theorem can be established for a Bayesian pure exchange economy with randomized choices in the same way as theorem 8.2.1. □

Chapter 11

Large Economy

Bayesian pure exchange economies

$$\mathcal{E}_{pe} := \left\{ C^j, T^j, u^j, e^j, \{\pi^j(\cdot \mid t^j)\}_{t^j \in T^j} \right\}_{j \in N}$$

considered in the previous chapter are economies with a finite population of consumers. This chapter reviews works dealing with core solutions within the framework of Bayesian pure exchange economy with an infinite population of consumers using the setup introduced by Aumann (1964). Propositions 11.2.3 and 11.3.4 are new.

11.1 Large Bayesian Pure Exchange Economy

Essentially all the works in the literature maintain finite information structures. They are formulated here as in a finite population case using finitely many type spaces, T^j, $j \in N$. (The set N is no longer a consumer set; it is an index set for different information structures.) Denote by T the type-profile space, $\prod_{j \in N} T^j$. A type profile may synonymously be called a state.

Let $(A, \mathcal{A}, \nu)$ be a measure space of consumers where ν represents an atomless population measure, i.e., $\nu(A) = 1$ and $\nu(S)$ is the proportion of consumers in the set or the coalition S ($\in \mathcal{A}$) to the totality of the population in A. There are l commodities, indexed by $h = 1, \ldots, l$. Each consumer $a \in A$ has the consumption set $\mathbf{R}^l_+$, his preference relation is represented by a type-profile-dependent von Neumann-Morgenstern utility function, $u(a, \cdot, \cdot) : \mathbf{R}^l_+ \times T \to \mathbf{R}$, and his initial endowment at type-profile t is a commodity bundle $e(a, t) \in \mathbf{R}^l_+$. For each $t \in T$, the map $u(\cdot, \cdot, t) : A \times \mathbf{R}^l_+ \to \mathbf{R}$, $(a, c) \mapsto u(a, c, t)$, is assumed to be $(\mathcal{A} \otimes \mathcal{B}^l) - \mathcal{B}$-measurable, where $\mathcal{B}^l$ and $\mathcal{B}$ are the Borel σ-algebra of subsets of $\mathbf{R}^l_+$ and

R respectively. For each $t \in T$, the function $e(\cdot, t)$ is assumed to be ν-integrable on A. Let $j(a)$ be the index for consumer a's private information structure; his type space is then $T^{j(a)}$. We assume that $j : A \to N$ is $\mathcal{A}$-measurable. For each coalition $S \in \mathcal{A}$ and index $i \in N$, the set S^i denotes $\{a \in S \mid j(a) = i\}$, and $j_+(S)$ denotes the set of all indices $i \in N$ such that $\nu(S^i) > 0$.

Each consumer $a \in A$ either has his *ex ante* (unconditional) probability on T or his *interim* (conditional) probability $\pi^a(\cdot \mid t^a)$ on $T^{N \setminus \{j(a)\}}$ given t^a.[1] Probability π^a may be objective or subjective. To sum up, a *large Bayesian pure exchange economy* is given by

$$\mathcal{E}_{PE} := \left((A, \mathcal{A}, \nu), \{\mathbf{R}_+^l, T^{j(a)}, u(a, \cdot, \cdot), e(a, \cdot), \pi^a\}_{a \in A} \right)$$

specified by the above conditions.

Let supp π^a be the support of π^a, $\{t \in T \mid \pi^a(\{t\}) > 0\}$, in case of an *ex ante* probability and be given by the set $T(\pi^a)$, as defined in the section 2.1, in case of an *interim* probability

The domain $T(S)$ of strategies of coalition S is *a priori* given so that

$$\bigcup_{a \in S'} \text{supp } \pi^a \subset T(S) \subset T$$

for some $S' \in \mathcal{A}$ for which $S' \subset S$ and $\nu(S \setminus S') = 0$. A feasible strategy bundle for coalition $S \in \mathcal{A}$ is a *feasible commodity allocation plan* for S, that is, a function $x : S \times T(S) \to \mathbf{R}_+^l$, such that the function $x(\cdot, t)$ is ν-integrable for each state t, and

$$\forall t \in T(S) : \int_S x(a,t)\nu(da) \leq \int_S e(a,t)\nu(da).$$

It is said to satisfy the *private measurability* condition if function $x(a, \cdot)$ is $\mathcal{T}^a$-measurable, ν-a.e. (see condition 3.1.1 for the null communication system).

11.2 *Interim* Solutions

We have seen Einy, Moreno and Shitovitz' (2000a) work in section 7.1 that a fine core allocation plan of a large Bayesian pure exchange economy $\mathcal{E}_{PE}$ is an ex post core allocation plan of $\mathcal{E}_{PE}$ (theorem 7.1.4), that this is not

---
[1] Notice that while there may be infinitely many different probabilities, π^a, $a \in A$, there are only finitely many different supports, due to the finiteness of T, so the equivalence of the two approaches discussed in section 2.3 still holds true.

necessarily true in the finite setup of $\mathcal{E}_{pe}$ (example 7.1.1), and that the converse is not true either.

The following result of Einy, Moreno and Shitovitz (2000b) gives a straightforward characterization of the *ex post* core which may not be private measurable nor Bayesian incentive compatible.

PROPOSITION 11.2.1 (Einy, Moreno and Shitovitz, 2000b) *Let $\mathcal{E}_{PE}$ be a large Bayesian pure exchange economy satisfying the following conditions:*
(i) *for every $t \in T(A)$, $e(a,t) \gg 0$, ν-a.e. ;*
(ii) *the function $u(a,\cdot,t): \mathbf{R}_+^l \to \mathbf{R}$ is continuous and strictly increasing on $\mathbf{R}_+^l$ for each $(a,t) \in A \times T$.*

Then, the ex post core of $\mathcal{E}_{PE}$ is nonempty and it is characterized by the set $CF(\mathcal{E}_{PE})$ defined as

$$\{x: A \times T(A) \to \mathbf{R}_+^l \mid x \text{ is a strategy bundle and } x(a,t) \text{ is}$$
$$\text{a core allocation of the large pure exchange economy}$$
$$\mathcal{E}_{PE}(t) \text{ for every } t \in T(A).\}.$$

Proof. We first show that the *ex post* core of $\mathcal{E}_{PE}$ is equal to $CF(\mathcal{E}_{PE})$. It is immediate that $CF(\mathcal{E}_{PE})$ is a subset of the *ex post* core of $\mathcal{E}_{PE}$. So, let x be a strategy bundle in the *ex post* core of $\mathcal{E}_{PE}$. Suppose $x \notin CF(\mathcal{E}_{PE})$; then, there exists $t_0 \in T(A)$ such that $x(a,t_0)$ is not a core allocation of the economy $\mathcal{E}_{PE}(t_0)$. Therefore, $(\exists\, S \in \mathcal{A} : \nu(S) > 0)$: $(\exists\, c: S \to \mathbf{R}_+^l : \nu$-integrable$)$ such that

$$\int_S c(a)\nu(da) \leq \int_S e(a,t_0)\nu(da), \text{ and}$$
$$u(a,c(a),t_0) > u(a,x(a,t_0),t_0), \nu\text{-a.e. in } S.$$

Define a map $y: S \times T(S) \to \mathbf{R}_+^l$ by

$$y(a,t) = \begin{cases} c(a) & \text{if } a \in S^j \text{ and } t^j = t_0^j \\ e(a,t) & \text{otherwise.} \end{cases}$$

Then, $y: S \times T(S) \to \mathbf{R}_+^l$ is a strategy bundle attainable in S,

$$(\forall t \in T(S)) \int_S y(a,t)\nu(da) \leq \int_S e(a,t)\nu(da),$$

and contradicts the property (ii) of the *ex post* core. Hence, x must be a strategy bundle in the set $CF(\mathcal{E}_{PE})$.

Next, let us check that the *ex post* core of $\mathcal{E}_{PE}$ is nonempty. By the standard arguments of Aumann (1964) and Aumann (1966) or Hildenbrand (1970), for each $t \in T(A)$ there is a core allocation $x_t : A \to \mathbf{R}_+^l$ of the economy $\mathcal{E}_{PE}(t)$. Define a map $x^* : A \times T(A) \to \mathbf{R}_+^l$ by $x^*(a,t) = x_t(a)$ for $a \in A$ and $t \in T(A)$. Then, x^* is a well-defined strategy bundle and $x^* \in CF(\mathcal{E}_{PE})$. □

We now want to extends the *interim* Bayesian incentive-compatible core (definition 5.1.5) to large Bayesian pure exchange economies. Suppose that the members of coalition S agree to take a strategy bundle $x : S \times T(S) \to \mathbf{R}_+^l$. Given a type profile $\bar{t} \in T$, consumer a's true type is $\bar{t}^{j(a)}$. If he makes a choice according to the agreement, his *interim* expected utility given his true type is

$$Eu\left(a, x(a, \cdot) \mid \bar{t}^{j(a)}\right) := \sum_{t \in T} u\left(a, x(a,t), t\right) \pi^a(t \mid \bar{t}^{j(a)}).$$

In the private information case, he can make any choice $c^a \in x(a, T^{j(a)}) \setminus \{x(a, \bar{t}^{j(a)})\}$ contrary to the agreement without being detected by his colleagues. If consumer a makes such a choice, his *interim* expected utility given his true type is

$$Eu\left(a, c^a \mid \bar{t}^{j(a)}\right) := \sum_{t \in T} u\left(a, c^a, t\right) \pi^a(t \mid \bar{t}^{j(a)}).$$

Thus, a strategy bundle $x : S \times T(S) \to \mathbf{R}_+^l$ is said to be *Bayesian incentive compatible* in a large Bayesian pure exchange economy if it satisfies:

$$\left(\forall t, \bar{t} \in T(S) : \ t^i = \bar{t}^i \text{ for } i \neq j(a)\right)$$
$$Eu^a\left(x(a,t) \mid t^j\right) \geq Eu^a\left(x(a,\bar{t}) \mid t^j\right), \ \nu\text{-a.e. in } S.$$

The following definition extends the *interim* Bayesian incentive-compatible core to large Bayesian pure exchange economies.

DEFINITION 11.2.2 A strategy bundle $x^* : A \times T(A) \to \mathbf{R}_+^l$ is said to be in the *interim Bayesian incentive-compatible core* if
(i) it is private measurable, Bayesian incentive compatible, and attainable in A,

$$\forall \ t \in T(A) : \int_A x^*(a,t)\nu(da) \leq \int_A e(a,t)\nu(da),$$

and
(ii) if it is not true that

$$(\exists \ S \in \mathcal{A} : \nu(S) > 0) : (\exists \ x : S \times T(S) \to \mathbf{R}_+^l : x(\cdot, t) \text{ is } \nu\text{-integrable and}$$

x is Bayesian incentive compatible) : $\forall\, t \in T(S)$:

$$\int_S x(a,t)\nu(da) \leq \int_S e(a,t)\nu(da),$$
$$Eu\left(a, x(a,\cdot) \mid t^{j(a)}\right) > Eu\left(a, x^*(a,\cdot) \mid t^{j(a)}\right)$$
$$\nu\text{-a.e. in } \{a \in S \mid j(a) \in j_+(S)\}.$$

Despite the fact that there are clear-cut conditions under which an *interim* Bayesian incentive-compatible core is nonempty (theorem 10.1.4) within the framework of Bayesian pure exchange economy, we remarked that the coalitional stability condition in the Bayesian incentive-compatible *interim* core is very strong. So much so that unless very specific conditions are given, we obtain non-existence results in many Bayesian pure exchange economies satisfying the "standard" assumptions as in the case of the example 10.1.3.

We show next that one of specific conditions under which the Bayesian incentive-compatible core of a large Bayesian pure exchange economy is nonempty is given by circumstances where every consumer knows his own utility and the endowment, that is, where utility functions and endowments are private measurable.

Here, as in the section 10.1 a strategy bundle for $S \in \mathcal{A}$ is a net-trade plan $z : S \times T(S) \to \mathbf{R}^l$ which is individually feasible, i.e., $z(a,t) + e(a,t) \geq 0$ for all $t \in T$, a.e. $a \in A$.

A large Bayesian pure exchange economy $\mathcal{E}_{PE} = ((A, \mathcal{A}, \nu), \{\mathbf{R}_+^l, T^{j(a)}, u(a,\cdot,\cdot), e(a,\cdot), \pi^a\}_{a \in A})$ is said to be of *finite utility-endowments-types*[2] if

$$(\forall i \in N)\, (\exists S \subset A^i = \{a \in A \mid j(a) = i\} : \nu(S) = \nu(A^i))$$
$$(\forall a, a' \in S)\, u(a,t) = u(a',t) \text{ and } e(a,t) = e(a',t),$$

and of *finite utility-types* if

$$(\forall i \in N)\, (\exists S \subset A^i = \{a \in A \mid j(a) = i\} : \nu(S) = \nu(A^i))$$
$$(\forall a, a' \in S)\, u(a,t) = u(a',t).$$

PROPOSITION 11.2.3 *Let $\mathcal{E}_{PE}$ be a large Bayesian pure exchange economy of finite utility-endowments-types satisfying the following conditions:*
(i) *for every $a \in A$, the map : $T \to \mathbf{R}_+^l$, $t \mapsto e(a,t)$, is $T^{j(a)} - \mathcal{B}^l$-measurable;*

---

[2] As long as there are only finitely many different utility functions and endowments the result below holds true. For simplicity, we assumed here that differences in utility functions and endowments come from a difference in consumer type $i \in N$.

(ii) *for every $c \in \mathbf{R}_+^l$ and $a \in A$ the map : $T \to \mathbf{R}$, $t \mapsto u(a,c,t)$, is $\mathcal{T}^{j(a)} - \mathcal{B}$-measurable;*
(iii) *for every $t \in T(A)$, $e(a,t) \gg 0$, ν-a.e.;*
(iv) *the function $u(a,\cdot,t) : \mathbf{R}_+^l \to \mathbf{R}$ is continuous, increasing, strictly quasi-concave on $\mathbf{R}_+^l$ for each $(a,t) \in A \times T$.*

Then, there exists an interim Bayesian incentive-compatible core net trade plan of the economy $\mathcal{E}_{PE}$.

Proof. Denote the restriction of $\mathcal{E}_{PE}$ to t, a large pure exchange economy with complete information, by

$$\mathcal{E}_{PE}(t) := \left((A, \mathcal{A}, \nu), \{\mathbf{R}_+^l, u(a,\cdot,t), e(a,t)\}_{a \in A}\right),$$

and define the set $CF(\mathcal{E}_{PE})$ as in the previous proposition 11.2.1.
We show first:

Claim: A strategy bundle $x^* : A \times T(A) \to \mathbf{R}_+^l$ in $CF(\mathcal{E}_{PE})$ satisfies the condition (ii) of the definition 11.2.2.

Suppose not; then,

$(\exists S \in \mathcal{A} : \nu(S) > 0) : (\exists x : S \times T(S) \to \mathbf{R}_+^l : x(\cdot,t)$ is ν-integrable$)$
$: \forall t \in T(S) :$

$$\int_S x(a,t)\nu(da) \leq \int_S e(a,t)\nu(da),$$
$$Eu\left(a, x(a,\cdot) \mid t^{j(a)}\right) > Eu\left(a, x^*(a,\cdot) \mid t^{j(a)}\right)$$
$$\nu\text{-a.e. in } \{a \in S \mid j(a) \in j_+(S)\}.$$

Since for every $c \in \mathbf{R}_+^l$ and $a \in A$ the map : $T \to \mathbf{R}$, $t \mapsto u(a,c,t)$, is $\mathcal{T}^{j(a)} - \mathcal{B}$-measurable, we have

$$Eu\left(a, x(a,\cdot) \mid t^{j(a)}\right) = u\left(a, x(a,\cdot), t^{j(a)}\right), \text{ and}$$
$$Eu\left(a, x^*(a,\cdot) \mid t^{j(a)}\right) = u\left(a, x^*(a,\cdot), t^{j(a)}\right),$$

for $t \in T(S)$ with $t^j = t^{j(a)}$, ν-a.e. in $\{a \in S \mid j(a) \in j_+(S)\}$. These contradict the fact that $x^* \in CF(\mathcal{E}_{PE})$. This proves the above claim.

It follows from the argument of Hildenbrand (1970) that for each $t \in T(A)$ a competitive net trade plan $z(\cdot,t) : A \to \mathbf{R}^l$ of large pure exchange economy with complete information $\mathcal{E}_{PE}(t)$ exists. The net trade plan as a map $z : A \times T(A) \to \mathbf{R}^l$ is a strategy bundle which is private measurable. Indeed, by assumptions (i) and (ii) of the proposition, it follows from the strict quasi-concavity of utility functions that for a.e. $a \in A$ we must have $z(a,t) = z(a,\bar{t})$ for any pair $t, \bar{t} \in T(A)$ with $t^{j(a)} = \bar{t}^{j(a)}$.

By the standard arguments of Aumann (1964) or Hildenbrand (1968) $z : A \times T(A) \to \mathbf{R}^l$ represents a core net trade plan $z(\cdot, t)$ for each $t \in T(A)$, and by what has been shown above the strategy bundle z is private measurable.

Let $i \in j_+(A)$. Then, since $\mathcal{E}_{PE}$ is of finite utility-endowments-types, it follows from the strict quasi-concavity of utility functions $u(a,t)$ for all $a \in A$ and $t \in T$ that $z(a,t)$'s are identical a.e. $a \in A^i$ for any $i \in j_+(A)$. Therefore, the The net trade plan z is symmetric in the sense that $(\forall i \in j_+(A))\,(\exists S \subset A^i = \{a \in A \mid j(a) = i\} : \nu(S) = \nu(A^i))\,(\forall a, a' \in S)\ z(a,t) = z(a',t)$. Since almost every consumer having the same type $i \in j_+(A)$ receives identical net trade bundle, let $z^i(t)$ denote this identical bundle for each $i \in j_+(A)$ and for each $t \in T(A)$. Then, we have

$$(\forall t \in T(A)) \int_A z(a,t)\nu(da) = \sum_{i \in j_+(A)} z^i(t)\nu(A^i) \leq 0.$$

Since $Eu(a, \cdot | t^{j(a)})$'s are monotone, it now follows from the argument of Ichiishi and Radner (lemma 9.2.10) that there exist private measurable mappings $\hat{z}^i : T(A) \to \mathbf{R}^l, i \in j_+(A)$, such that

$$(\forall t \in T(A))(\forall i \in j_+(A))\ z^i(t) \leq \hat{z}^i(t), \text{ and}$$
$$(\forall t \in T(A)) \sum_{i \in j_+(A)} \hat{z}^i(t)\nu(A^i) = 0.$$

By Hahn and Yannelis' proposition (proposition 4.1.3 of this book), it is Bayesian incentive compatible. Define a strategy bundle $z^* : A \times T(A) \to \mathbf{R}^l$ by

$$(\forall i \in j_+(A))(\forall a \in A^i)(\forall t \in T(A))\ z^*(a,t) := \hat{z}^i(t), \text{ and}$$
$$(\forall a \notin \cup_{i \in j_+(A)} A^i)(\forall t \in T(A))\ z^*(a,t) := z(a,t).$$

Then, the bundle $z^* : A \times T(A) \to \mathbf{R}^l$ is an *interim* Bayesian incentive-compatible core net trade plan of the large Bayesian pure exchange economy $\mathcal{E}_{PE}$. □

11.3 *Ex Ante* Solutions

A *state-contingent claim* for commodity h is a commodity traded in the *ex ante* period which promises delivery of a unit of commodity h upon realization of information state t in the *interim* period, and no delivery upon realization of any other state. A *claim allocation* in A is a function $x : A \times T \to \mathbf{R}_+^l$, assigning to each consumer a the claim bundle $x(a, \cdot)$, which is attainable in the economy,

$$\forall\, t \in T : \int_A x(a,t)\nu(da) \le \int_A e(a,t)\nu(da).$$

DEFINITION 11.3.1 A strategy bundle $x^* : A \times T(A) \to \mathbf{R}_+^l$ is said to be in the *ex ante private information core* of a large Bayesian pure exchange economy if
(i) it is private measurable claim allocation in A, and
(ii) if it is not true that

$$(\exists\, S \in \mathcal{A} : \nu(S) > 0) : (\exists\, x : S \times T(S) \to \mathbf{R}_+^l : x \text{ is private measurable}$$
and $x(\cdot, t)$ is ν-integrable$) : \forall\, t \in T(S) :$
$$\int_S x(a,t)\nu(da) \le \int_S e(a,t)\nu(da),$$
$$Eu(a, x(a, \cdot)) > Eu(a, x^*(a, \cdot)),$$
$$\nu\text{-a.e. in } \{a \in S \mid j(a) \in j_+(S)\}.$$

Einy, Moreno and Shitovitz (2001a) studied an *ex ante* private information core, in a large Bayesian pure exchange economy $\mathcal{E}_{PE}$ with a finite *ex ante* state space. Here we describe their result using the type-profile space.

Their idea and methods are straightforward. They consider Radner's (1968) *ex ante competitive equilibrium* of the state-contingent claim market defined as a pair (p^*, x^*) of price vector $p^* : T \to \mathbf{R}_+^l$ and measurable claim allocation $x^* : A \times T(A) \to \mathbf{R}_+^l$ such that ν-a.e., consumer a's claim bundle $x^*(a, \cdot)$ maximizes his *ex ante* expected utility subject to the budget constraint:

$$\text{Maximize} \quad \sum_{t \in T} \pi^a(t) u(a, x(a,t), t)$$
$$\text{subject to} \quad x(a, \cdot) \text{ is } \mathcal{T}^a\text{-measurable},$$
$$\sum_{t \in T} p^*(t) \cdot x(a,t) \le \sum_{t \in T} p^*(t) \cdot e(a,t),$$

and Walras' law is satisfied with equality,

$$\sum_{t \in T} p^*(t) \cdot \int_A x^*(a,t)\nu(da) = \sum_{t \in T} p^*(t) \cdot \int_A e(a,t)\nu(da).$$

Applying the standard arguments, Einy, Moreno and Shitovitz established the following theorem:

PROPOSITION 11.3.2 (Einy, Moreno and Shitovitz, 2001a) *Suppose economy $\mathcal{E}_{PE}$ satisfies: For ν-a.e. in A and for every $t \in T$,*
(i) $e(a,t) \gg 0$; and
(ii) *the function $u(a,\cdot,t) : \mathbf{R}_+^l \to \mathbf{R}$ is continuous and monotone.*
Then, the ex ante private information core of $\mathcal{E}_{PE}$ is nonempty, and it is identical to the set of ex ante competitive allocations.

Proof. It follows from the arguments of Aumann (1964) or Hildenbrand (1968) that the set of *ex ante* competitive allocations is identical to the *ex ante* private information core. Using the argument of Hildenbrand (1970, Theorem 2), one can show that an *ex ante* competitive allocation exists. □

The above proposition clarifies the fact that the basic results for large economies with perfect information still holds for *ex ante* private core and *ex ante* competitive allocations in large Bayesian pure exchange economies. However, there is one essential drawback in this result. As was discussed in Chapter 4, since Bayesian incentive compatibility conditions need not be satisfied by private core allocations, feasibility of execution of agreements of a strategy bundle may not be warranted.

DEFINITION 11.3.3 A strategy bundle $x^* : A \times T(A) \to \mathbf{R}_+^l$ is said to be in the *ex ante Bayesian incentive compatible core* of a large Bayesian pure exchange economy if
(i) it is a private measurable and Bayesian incentive compatible claim allocation in A,
(ii) if it is not true that

$(\exists\, S \in \mathcal{A} : \nu(S) > 0) : (\exists\, x : S \times T(S) \to \mathbf{R}_+^l : x(\cdot,t)$ is ν-integrable and x is Bayesian incentive compatible $) : \forall\, t \in T(S) :$

$$\int_S x(a,t)\nu(da) \leq \int_S e(a,t)\nu(da),$$
$$Eu(a, x(a,\cdot)) > Eu(a, x^*(a,\cdot)),$$

ν-a.e. in $\{a \in S \mid j(a) \in j_+(S)\}$.

We present here a positive existence result on the *ex ante* Bayesian incentive compatible core.

PROPOSITION 11.3.4 *Let $\mathcal{E}_{PE}$ be a large Bayesian pure exchange economy of finite utility-endowments-types satisfying the following conditions:*
(i) *for every $t \in T(A)$, the map $e(\cdot,t) : A \to \mathbf{R}_+^l$ is ν-integrable and $e(a,t) \gg 0$ for almost every $a \in A$;*

(ii) *the function $u(a,\cdot,t) : \mathbf{R}_+^l \to \mathbf{R}$ is continuous, monotone, and strictly quasi-concave for each $(a,t) \in A \times T$.*

Then there exists an ex ante Bayesian incentive-compatible core net-trade plan.

Proof. Using the argument of Hildenbrand (1970), one can show that an *ex ante* competitive net trade plan exists. It then follows from the arguments of Aumann (1964) or Hildenbrand (1968) that an *ex ante* private information core net trade plan exists since the set of *ex ante* competitive allocations is identical to the *ex ante* private information core.

Let $z : A \times T(A) \to \mathbf{R}_+^l$ be such an *ex ante* private information core net trade plan which is an *ex ante* competitive net trade plan. Let $i \in j_+(A)$. Then, since $\mathcal{E}_{PE}$ is of finite utility-endowments-types, it follows from the strict quasi-concavity of utility functions $u(a,t)$ for all $a \in A$ and $t \in T$ that $z(a,t)$'s are identical a.e. $a \in A^i$ for any $i \in j_+(A)$). Therefore, the *ex ante* private information core net trade plan z is symmetric in the sense that $(\forall i \in j_+(A)) \left(\exists S \subset A^i = \{a \in A \mid j(a) = i\} : \nu(S) = \nu(A^i)\right)(\forall a, a' \in S)\ z(a,t) = z(a',t)$. Since almost every consumers having the same type $i \in j_+(A)$ receive identical net trade bundle, let $z^i(t)$ denote this identical bundle for each $i \in j_+(A)$ and for each $t \in T(A)$. Then, we have

$$(\forall t \in T(A)) \int_A z(a,t)\nu(da) = \sum_{i \in j_+(A)} z^i(t)\nu(A^i) \leq 0.$$

Since $Eu(a,\cdot)$'s are monotone, it now follows from the argument of Ichiishi and Radner (lemma 9.2.10) that there exist private measurable mappings $\hat{z}^i : T(A) \to \mathbf{R}_+^l, i \in j_+(A)$, such that

$$(\forall t \in T(A))(\forall i \in j_+(A))\ z^i(t) \leq \hat{z}^i(t),\ \text{and}$$
$$(\forall t \in T(A)) \sum_{i \in j_+(A)} \hat{z}^i(t)\nu(A^i) = 0.$$

By Hahn and Yannelis' proposition (proposition 4.1.3 of this book), it is Bayesian incentive compatible. Define a strategy bundle $z^* : A \times T(A) \to \mathbf{R}_+^l$ by

$$(\forall i \in j_+(A))(\forall a \in A^i)(\forall t \in T(A))\ z^*(a,t) := \hat{z}^i(t),\ \text{and}$$
$$(\forall a \notin \cup_{i \in j_+(A)} A^i)(\forall t \in T(A))\ z^*(a,t) := z(a,t).$$

Then, the bundle $z^* : A \times T(A) \to \mathbf{R}_+^l$ is the required *ex ante* Bayesian incentive-compatible core strategy bundle of the large Bayesian pure exchange economy $\mathcal{E}_{PE}$. □

Beth Allen (1999) considered the existence of *ex ante* Bayesian incentive-compatible core of a large Bayesian pure exchange economy $\mathcal{E}_{PE}$ with finite utility-types but an infinite variation of endowments. As a matter of fact, her result hinges on dispersion of wealth distribution induced by the endowments distribution of consumers as in the works of a general equilibrium existence theorem in Yamazaki (1978, 1981). Moreover, there is an additional cost of allowing an infinite variations of endowments among consumers in terms of tighter restrictions on the class of utility functions.

Given a large Bayesian pure exchange economy of finite utility-types $\mathcal{E}_{PE}$, let C denote the endowment distribution among type $i \in j_+(A)$ in the economy $\mathcal{E}_{PE}$. Denote by λ_p^i the wealth distribution induced by η^i at a price vector $p \in \mathbf{R}_+^l$, that is,

$$\lambda_p^i(B) := \eta^i\left(\{e \in \mathbf{R}_{++}^l | p \cdot e \in B\}\right) \text{ for Borel subsets } B \subset \mathbf{R}_+.$$

PROPOSITION 11.3.5 (Allen, 1999) *Let $\mathcal{E}_{PE}$ be a large Bayesian pure exchange economy of finite utility-types satisfying the following conditions: For ν-a.e. in A and for every $t \in T$,*
(i) [Differentiable strict monotonicity] *the function $u(a, \cdot, t) : \mathbf{R}_+^l \to \mathbf{R}$ is C^1 on $\mathbf{R}_{++}^l$ and $Du(a, y, t) \gg 0$ for all $y \in \mathbf{R}_{++}^l$.*
(ii) [Boundary condition for indifference curves] *for each $i \in j_+(A)$, the support $\mathrm{supp}\eta^i$ of η^i is compact and*

$$\mathrm{cl}\left(\left\{y \in \mathbf{R}_{++}^{\sharp T(A) \times l} \mid Eu(a, e, \cdot) \leq Eu(a, y(t), \cdot)\right\}\right) \bigcap \partial \mathbf{R}_+^l = \emptyset.$$

for almost every $e \in \mathbf{R}_{++}^l$ [3], and
(iii) [Full rank condition] *$l \geq \sharp T(A)$ and the matrix*

$$\left(Du\left(a, y(t^1), t^1\right), \ldots, Du\left(a, y(t^{\sharp T(A)}), t^{\sharp T(A)}\right)\right)$$

has full rank almost everywhere in $\mathbf{R}_{++}^{l \times \sharp T(A)}$.

Then there exists an ex ante Bayesian incentive-compatible core net-trade plan.

Proof. See B. Allen (Proposition 5.4 and Corollary 6.8, 1999). □

The basic idea of the proof is to note that the possible nonexistence of *ex ante* Bayesian incentive-compatible core allocation is due to nonconvexity arising from the Bayesian incentive-compatibility condition. Thus Allen pointed out that this difficulty can be overcome just as in the case of general

---
[3] With respect to the l-dimensional Lebesgue measure.

equilibrium existence problem of a large economy when consumption sets and/or the underlying commodity space itself may not satisfy convexity (see Mas-Colell (1977), and Yamazaki (1978, 1981)). Stronger requirements on utility functions are to ensure local nonsatiation condition for strategy bundles satisfying the Bayesian incentive-compatibility condition.

The above two propositions that established the existence of an *ex ante* Bayesian incentive-compatible core strategy bundle rely on the finite utility-types. Whether its existence can be established with an infinite utility types is still an open question.

Chapter 12

Core Convergence/Equivalence Theorems

We turn to a bulk of works on the Edgeworth conjecture (the core convergence theorem or the core equivalence theorem) for the Bayesian pure exchange economy. In the course of studying this issue, various competitive equilibrium concepts have been invoked or newly proposed; some suffer from conceptual difficulties. We believe, however, that a core convergence/equivalence result is meaningful only if it approximates/characterizes a competitive equilibrium which is defined sensibly enough so that one expects to realize in the competitive market.

12.1 *Interim* Solutions

We first present Serrano, Vohra and Volij's (2001) negative result on Wilson's (1978) coarse core allocations (definition 5.1.1) in the Bayesian pure exchange economy,

$$\mathcal{E}_{pe} := \left\{ \mathbf{R}_+^l, T^j, u^j, e^j, \{\pi^j(\cdot \mid t^j)\}_{t^j \in T^j} \right\}_{j \in N},$$

replicated as in Debreu and Scarf (1963); they did not impose private measurability (condition 3.1.1 for the null communication system) or Bayesian incentive compatibility (condition 4.1.1 or 4.2.1).

Here, we have to be careful in defining the term *replication*. Let $n := \#N$. The q-replica economy of $\mathcal{E}_{pe}$ is meant to consist of $n \cdot q$ consumers, indexed by $N \times \{1, 2, \ldots, q\}$, so that consumer $(j, k) \in N \times \{1, 2, \ldots, q\}$ has the same consumer characteristics as consumer j of $\mathcal{E}_{pe}$. But in the economy with $n \cdot q$ consumers, the type-profile space should be the set

$$\overbrace{T \times T \times \ldots \times T}^{q},$$

so consumer (j, k)'s type-profile-dependent von Neumann-Morgenstern utility function should be defined on $\mathbf{R}_+^l \times \overbrace{T \times T \times \ldots \times T}^{q}$, and his *interim* probability given his type should be defined on $T^{N\setminus\{j\}} \times \overbrace{T \times T \times \ldots \times T}^{q-1}$. Therefore, consumer characteristics become different for different numbers of replication.

Serrano, Vohra and Volij postulated for the q-replica economy that the types of the consumers (j, k), $k = 1, \ldots, q$, are all described by the same set T^j, that the types of (j, k) and of (j, k') are perfectly correlated, and that their *interim* probabilities are the same as $\pi^j(\cdot \mid t^j)$. This idea is precisely formulated as follows: The *interim* probability of player (j, k) given his type $\bar{t}^j$, $\pi^{(j,k)}(\cdot \mid \bar{t}^j)$, concentrates on the "diagonal" of the space

$$T^{N\setminus\{j\}} \times \overbrace{\left(\{\bar{t}^j\} \times T^{N\setminus\{j\}}\right) \times \ldots \times \left(\{\bar{t}^j\} \times T^{N\setminus\{j\}}\right)}^{q-1},$$

and is given as

$$\pi^{(j,k)}\left(t^{N\setminus\{j\}}, (\bar{t}^j, t^{N\setminus\{j\}}), \ldots, (\bar{t}^j, t^{N\setminus\{j\}}) \mid \bar{t}^j\right) := \pi^j(t^{N\setminus\{j\}} \mid \bar{t}^j).$$

The diagonal is identified with $\{\bar{t}^j\} \times T^{N\setminus\{j\}}$. Thus, the same state space $\Omega := T$ is given to each q-replica economy, and player (j, k)'s information structure is formulated by the spaces, $(T, T^j, \pi^j(\cdot \mid t^j))$, $t^j \in T^j$. Player (j, k)'s utility function and initial endowment function are identical to those of player j of the original economy $\mathcal{E}_{pe}$, that is, functions $u^j : \mathbf{R}_+^l \times T \to \mathbf{R}$ and $e^j : T^j \to \mathbf{R}_+^l$.

We assume for simplicity

$$T(S) = T(S') =: T^* \subset T, \text{ for all } S, S' \in \mathcal{N}.$$

Recall supp $\pi^j \subset T^*$.

As a competitive equilibrium concept for $\mathcal{E}_{pe}$ which belongs to the coarse core, Serrano, Vohra and Volij followed Wilson (1978, footnote 6, page 814) and considered the following constrained market equilibrium concept:

Let $\Delta := \{p : T^* \to \mathbf{R}_+^l \mid \sum_{t \in T} \sum_{h=1}^{l} p_h(t) = 1\}$ be the price domain. Consumer j's consumption plan is a strategy $x^j : T^* \to \mathbf{R}_+^l$. A consumption plan x^j gives rise to the *interim* expected utility given type $\bar{t}^j$, $Eu^j(x^j \mid \bar{t}^j) := \sum_{t \in T} u^j(x^j(t), t) \pi^j(t \mid \bar{t}^j)$. An allocation is a consumption plan bundle x such that the total demand is equal to the total supply at every type profile,

$$\forall\, t \in T^* : \sum_{j \in N} x^j(t) = \sum_{j \in N} e^j(t^j).$$

A *constrained market equilibrium* is a pair (p^*, x^*) of price vector p^* and allocation x^* such that for each consumer j and each information $\bar{t}^j$, x^{*j} maximizes his conditional expected utility given $\bar{t}^j$ subject to the budget constraint given $\bar{t}^j$:

Maximize $\quad Eu^j(x^j \mid \bar{t}^j)$

subject to $\quad \sum_{t^{N \setminus \{j\}}} p(\bar{t}^j, t^{N \setminus \{j\}}) \cdot x^j(\bar{t}^j, t^{N \setminus \{j\}})$

$\leq \sum_{t^{N \setminus \{j\}}} p(\bar{t}^j, t^{N \setminus \{j\}}) \cdot e^j(\bar{t}^j, t^{N \setminus \{j\}}).$

We will comment on this equilibrium concept later (remark 12.1.2).

Serrano, Vohra and Volij first noted the easy result that *a constrained market equilibrium allocation is a coarse core allocation*. Their main result is the following counterexample, which claims that the coarse core does not necessarily convergence to the constrained market equilibria as $q \to \infty$.

EXAMPLE 12.1.1 (Serrano, Vohra and Volij, 2001) Consider the following two-person, two-commodity, two-state Bayesian pure exchange economy $\mathcal{E}_{pe}$: Consumer 1 is informed, and consumer 2 is uninformed,

$$N = \{1, 2\}, \quad T^1 = \{l, h\}, \quad T^2 = \{t^2\},$$

so T is identified with T^1.

$$u^j(c_1, c_2, t) = (c_1 \cdot c_2)^{\frac{1}{4}}, \quad \text{for all } j \in N,\ c \in \mathbf{R}_+^2 \text{ and } t \in T;$$

$$e^1(t) = \begin{pmatrix} 24 \\ 0 \end{pmatrix}, \quad \text{for all } t \in T;$$

$$e^2(t) = \begin{pmatrix} 0 \\ 24 \end{pmatrix}, \quad \text{for all } t \in T;$$

$$\pi^2(t) = \frac{1}{2}, \quad \text{for all } t \in T.$$

Serrano, Vohra and Volij showed that this economy has a particular allocation x whose q-replication is in the coarse core of the q-replica economy for all q, yet x cannot be a constrained market equilibrium allocation.

Indeed, it is easy to show that the unique constrained competitive equilibrium (p^*, x^*) is given by

$$p^*(t) = \begin{pmatrix} \frac{1}{4} \\ \frac{1}{4} \end{pmatrix}, \quad x^{*j}(t) = \begin{pmatrix} 12 \\ 12 \end{pmatrix}, \quad \text{for all } j \in N, \text{ and all } t \in T.$$

On the other hand, define allocation x by

$$x^1(l) = \begin{pmatrix} 15 \\ 15 \end{pmatrix}, \quad x^1(h) = \begin{pmatrix} 8 \\ 8 \end{pmatrix}, \quad x^2(l) = \begin{pmatrix} 9 \\ 9 \end{pmatrix}, \quad x^2(h) = \begin{pmatrix} 16 \\ 16 \end{pmatrix}.$$

It suffices to show that the q-replication of x is in the coarse core of the q-replica economy for all q. Suppose the contrary, i.e., that there are q and coalition S in the q-replica economy, such that S improves upon x by using its allocation $y : T \to \left(\mathbf{R}_+^2\right)^S$. Without loss of generality, we may assume that y satisfies the equal treatment property (so $y = (y^1, y^2)$, $y^j : T \to \mathbf{R}_+^2$, $j \in N$) and is coarse efficient in S (definition 6.1.1 in which S replaces N). Let k_j be the number of players in S who have the same characteristics as player j of $\mathcal{E}_{pe}$, $k_j := \#\{(j,k) \in S \mid 1 \leq k \leq q\}$, $j \in N$. Coarse blocking means:

$$u^1(y^1(l)) > u^1(x^1(l)) = \sqrt{15}, \quad (12.1)$$
$$u^1(y^1(h)) > u^1(x^1(h)) = \sqrt{8}, \quad (12.2)$$
$$u^2(y^2(l)) + u^2(y^2(h)) > u^2(x^2(l)) + u^2(x^2(h)) = 7. \quad (12.3)$$

By the individual rationality of x, it follows that $k_j > 0$, $j \in N$.

The *interim* efficiency of y in S means the equality of the marginal rates of substitution across the consumers for each state,

$$\frac{\left.\frac{\partial u^1}{\partial c_1^1}\right|_{y^1(l)}}{\left.\frac{\partial u^1}{\partial c_2^1}\right|_{y^1(l)}} = \frac{\left.\frac{\partial u^2}{\partial c_1^2}\right|_{y^2(l)}}{\left.\frac{\partial u^2}{\partial c_2^2}\right|_{y^2(l)}}, \quad \frac{\left.\frac{\partial u^1}{\partial c_1^1}\right|_{y^1(h)}}{\left.\frac{\partial u^1}{\partial c_2^1}\right|_{y^1(h)}} = \frac{\left.\frac{\partial u^2}{\partial c_1^2}\right|_{y^2(h)}}{\left.\frac{\partial u^2}{\partial c_2^2}\right|_{y^2(h)}},$$

which implies under the Cobb-Douglas utility function that the commodity ratios across the consumers are the same for each state,

$$\frac{y_2^1(l)}{y_1^1(l)} = \frac{y_2^2(l)}{y_1^2(l)}, \quad \frac{y_2^1(h)}{y_1^1(h)} = \frac{y_2^2(h)}{y_1^2(h)}.$$

In the light of the market clearance condition, these identities imply that $y^1(t)$ and $y^2(t)$ are both proportional to $e^{(S)}(t) := k_1 e^1(t) + k_2 e^2(t)$, $t \in T$.

Thus,

$$\exists\, \alpha \in (0,1): \quad k_1 y^1(l) = \alpha e^{(S)}(l) = \alpha \begin{pmatrix} 24k_1 \\ 24k_2 \end{pmatrix},$$

$$k_2 y^2(l) = (1-\alpha) e^{(S)}(l) = (1-\alpha) \begin{pmatrix} 24k_1 \\ 24k_2 \end{pmatrix},$$

$$\exists\, \beta \in (0,1): \quad k_1 y^1(h) = \beta e^{(S)}(h) = \beta \begin{pmatrix} 24k_1 \\ 24k_2 \end{pmatrix},$$

$$k_2 y^2(h) = (1-\beta) e^{(S)}(h) = (1-\beta) \begin{pmatrix} 24k_1 \\ 24k_2 \end{pmatrix}.$$

Substituting these into the utility functions,

$$\begin{aligned}
k_1 \left(u^1(y^1(l)) \right)^2 &= 24\alpha \sqrt{k_1 k_2}, \\
k_2 \left(u^2(y^2(l)) \right)^2 &= 24(1-\alpha) \sqrt{k_1 k_2}, \\
k_1 \left(u^1(y^1(h)) \right)^2 &= 24\beta \sqrt{k_1 k_2}, \\
k_2 \left(u^2(y^2(h)) \right)^2 &= 24(1-\beta) \sqrt{k_1 k_2}.
\end{aligned}$$

Therefore, letting $z := k_1/k_2$,

$$z \left(u^1(y^1(t)) \right)^2 + \left(u^2(y^2(t)) \right)^2 = 24\sqrt{z}, \text{ for each } t \in T.$$

The inequalities (12.1) and (12.2) are re-written as

$$\begin{aligned}
\left(u^2(y^2(l)) \right)^2 &< 24\sqrt{z} - 15z, \\
\left(u^2(y^2(h)) \right)^2 &< 24\sqrt{z} - 8z.
\end{aligned}$$

By (12.3),

$$\sqrt{24\sqrt{z} - 15z} + \sqrt{24\sqrt{z} - 8z} > 7,$$

But this is impossible, since it is easily seen that the left-hand side achieves its maximum of 7 at $z = 1$. □

REMARK 12.1.2 Apart from failing to satisfy the measurability requirement, a constrained market equilibrium differs from Radner's (1968) *ex ante* competitive equilibrium of the state-contingent claim market in that it accommodates $\#T^j$ constrained maximization problems that each consumer j possibly faces at the *interim* period. It raises the following serious conceptual questions: Player j, acting alone in the market in which everybody is anonymous, and knowing that his true type is $\bar{t}^j$, does not bother acting rationally in the unrealized event $E := \{t \in T^* \mid t^j \neq \bar{t}^j\}$, yet his actions in E influence the competitive equilibrium price vector p^*. A more serious problem is that there is no reason why he should segment the market into $\#T^j$ submarkets. This point is all the more problematic since the way to segment the market differs among different anonymous consumers (commodity $(h, t^i, t^j, t^{N\setminus\{i,j\}})$ is traded with commodity $(k, t'^i, t^j, t^{N\setminus\{j\}})$ in consumer j's mind, yet they cannot be traded in consumer i's mind). $\square$

We turn to the fine core. We have pointed out that Einy, Moreno and Shitovitz (2000a) established within the framework of a nonatomic space of consumers,

$$\mathcal{E}_{PE} := \left((A, \mathcal{A}, \nu), \{\mathbf{R}_+^l, T^{j(a)}, u(a, \cdot, \cdot), e(a, \cdot), \pi^a\}_{a \in A} \right),$$

that a fine core allocation is an *ex post* core allocation (theorem 7.1.4 of this book). Notice that the *ex post* stage is essentially the complete information stage. They invoked Aumann's (1964) equivalence theorem for the pure exchange economy with complete information, and asserted the following corollary:

COROLLARY 12.1.3 (Einy, Moreno and Shitovitz, 2000a) *Let $\mathcal{E}_{PE}$ be a Bayesian pure exchange economy with a nonatomic positive measure space of consumers $(A, \mathcal{A}, \nu)$ and finitely many types $\{T^j\}_{j \in N}$, such that the domains of strategies for coalitions S satisfy*

$$T(S) \subset T(Q), \text{ for all } Q \in \mathcal{A} \text{ for which } j_+(Q) = N,$$

where $j(a)$ is the index for consumer a's private information structure, and for any $S \in \mathcal{A}$, $j_+(S)$ is the set of all indeces $i \in N$ for which $\nu(\{a \in S \mid j(a) = i\}) > 0$. Assume:
(i) *$j_+(A) = N$, that is, $\nu(A^i) > 0$ for every $i \in N$, where $A^i := \{a \in A \mid j(a) = i\}$;*
(ii) *for every $t \in T$, the map: $A \times \mathbf{R}_+^l \to \mathbf{R}$, $(a, c) \mapsto u(a, c, t)$, is $\mathcal{A} \otimes \mathcal{B}$-measurable, where $\mathcal{B}$ is the Borel σ-algebra of subsets of $\mathbf{R}_+^l$;*
(iii) *$\forall\, t \in T(A) : \int_A e(a, t) \gg 0$;*
(iv) *either the function $u(a, \cdot, t) : \mathbf{R}_+^l \to \mathbf{R}$ is continuous and strictly increasing for each $(a, t) \in A \times T$, or it is continuous, increasing, and vanishes on the boundary of $\mathbf{R}_+^l$ for each $(a, t) \in A \times T$.*
Then, a fine core allocation of $\mathcal{E}_{PE}$ is an ex post competitive allocation.

12.2 *Ex Ante* Solutions

We have seen in section 11.3 that Einy, Moreno and Shitovitz (2001a) studied an *ex ante* private measurable core allocation (a private core allocation of definition 5.2.1), in the large Bayesian pure exchange economy,

$$\mathcal{E}_{PE} := \left((A, \mathcal{A}, \nu), (\Omega, \mathcal{T}, \pi), \{\mathbf{R}_+^l, u(a,\cdot,\cdot), e(a,\cdot), \mathcal{T}^a\}_{a \in A} \right),$$

with a nonatmic measure space of consumers $(A, \mathcal{A}, \nu)$ and a finite *ex ante* state space $(\Omega, \mathcal{T}, \pi)$, $\#\Omega < \infty$, in which each consumer a has a private information structure as a subalgebra $\mathcal{T}^a$ of $\mathcal{T}$. Their result is that *an ex ante competitive equilibrium of the state-contingent market exists, and the set of ex ante competitive allocations is identical to the set of private information core allocations.* (proposition 11.3.2).

Forges, Heifetz and Minelli (2001) studied the Bayesian pure exchange economy,

$$\mathcal{E}_{pe} := \{\mathbf{R}_+^l, T^j, u^j, e^j, \pi\}_{j \in N},$$

in which each initial endowment is a constant function, $T \to \mathbf{R}^l$, $t \mapsto e^j$, and in which everybody j in coalition S chooses a mixed strategy (or rather, type-profile dependent mixed choices) $\mu^j : T^S \to \mathcal{M}(C^j)$, and the coalitional attainability is defined as the *ex ante* expected feasibility,

$$\sum_{j \in S} \sum_{t^S \in T^S} \pi(\{t^S\} \times T^{N \setminus S}) \int_{C^j} c^j \mu^j[t^S](dc^j) \leq \sum_{j \in S} e^j,$$

that is, the market clearance on the average across type profiles as well as across pure choices (compare with the attainability definition in corollary 10.2.1 as the expected feasibility at each type profile). Notice that the *ex ante* expected feasibility does not guarantee the expected feasibility at the *interim* time of strategy execution. They took the mediator-based approach, and considered the Bayesian incentive compatibility (see condition 4.2.1),

$$\forall \bar{t}^j, \tilde{t}^j \in T^j :$$

$$\sum_{t \in T} \pi(t \mid \bar{t}^j) U^j(\mu^j[t^S], t) \geq \sum_{t \in T} \pi(t \mid \bar{t}^j) U^j(\mu^j[\tilde{t}^j, t^{S \setminus \{j\}}], t),$$

where

$$U^j(\mu^j[\tilde{t}^j, t^{S \setminus \{j\}}], t) := \int_{C^j} u^j(c^j, t) \mu^j[\tilde{t}^j, t^{S \setminus \{j\}}](dc^j).$$

Denote by $F^{ic,S}$ the set of all strategy bundles of coalition S that satisfy the *ex ante* expected feasibility and the Bayesian incentive compatibility. Set for simplicity, $F^{ic} := F^{ic,N}$. The *ex ante* expected utility of strategy μ^j is given as $EU^j(\mu^j) := \sum_{t \in T} \pi(t) U^j(\mu^j[t^S], t)$.

Forges, Heifetz and Minelli defined an *ex ante core allocation* as a strategy bundle $\{\mu^{*j}\}_{j \in N}$, such that (i) it is attainable in the grand coalition,

$$\{\mu^{*j}\}_{j \in N} \in F^{ic},$$

and (ii) it is not weakly improved upon by any coalition,

$$\neg \, \exists \, S \in \mathcal{N} : \exists \, \{\mu^j\}_{j \in S} \in F^{ic,S} : \quad \forall \, j \in S : \quad EU^j(\mu^j) \geq EU^j(\mu^{*j}),$$
$$\exists \, j \in S : \quad EU^j(\mu^j) > EU^j(\mu^{*j}).$$

They defined an *ex ante competitive equilibrium* as a pair of a price vector $p^* \in \mathbf{R}_+^l$ and an attainable mixed-consumption plan bundle (strategy bundle) $\{\mu^{*j}\}_{j \in N} \in F^{ic}$, such that each mixed commodity bundle μ^{*j} satisfies the *ex ante* expected budget constraint,

$$\sum_{t \in T} \pi(t) \int_{C^j} \sum_{h=1}^{l} p_h c_h^j \mu^{*j}[t](dc^j) \leq \sum_{h=1}^{l} p_h e_h^j,$$

and is the best of those mixed commodity bundles satisfying the *ex ante* expected budget constraint, that is, for any Bayesian incentive compatible mixed commodity bundle μ^j for which $\sum_{t \in T} \pi(t) \int_{C^j} \sum_{h=1}^{l} p_h c_h^j \mu^j[t](dc^j) \leq \sum_{h=1}^{l} p_h e_h^j$, it follows that

$$EU^j(\mu^j) \leq EU^j(\mu^{*j}).$$

Applying the standard argument, Forges, Heifetz and Minelli established the following theorem:

THEOREM 12.2.1 (Forges, Heifetz and Minelli, 2001) *There exists an ex ante competitive equilibrium, and each ex ante competitive allocation is an ex ante core allocation.*

Then they considered the replica economies à la Debreu and Scarf (1963), in which the type profile space in the q-replica economy is the q-fold product of space T and the *ex ante* probability is the q-fold product probability of π. They established the following theorem:

THEOREM 12.2.2 (Forges, Heifetz and Minelli, 2001) (i) *When utility function u^j depends fully on the type profile t, there is an example in which the equal-treatment property is not valid for an ex ante core allocation of the q-replica economy.*
(ii) *In the case of no externalities ($u^j = u^j(c^j, t^j)$), the q-replica economy has an ex ante core allocation with the equal treatment property, and a consumption plan bundle with the equal treatment property which is in the ex ante core of all q-replica economies, $q = 1, 2, \cdots$, is an ex ante competitive allocation.*

REMARK 12.2.3 We have already pointed out the weakness of the attainability condition in their definition of *ex ante* core allocation. More serious problems show up in their mediator-based approach to the *ex ante* competitive allocation: each mixed commodity bundle μ^j depends fully on T, so it is not clear how a consumer can choose his mixed demand contingent upon the others' private information while being uncommunicative, and the *ex ante* notions of attainability and budget constraint fail to guarantee the attainability and the budget constraint at the *interim* time of actually executing these consumption plans. More importantly, they have not provided a rationale for imposing Bayesian incentive compatibility on the competitive allocations. A competitive equilibrium is an outcome of a specific noncooperative behavior guided only by a price vector established in the market, each consumer chooses his (mixed) commodity bundle by himself without coordinating with other consumers, so there is no need for him to promise truthful execution of his strategy to anybody. Thus, the sensible setup would be that an *ex ante* core allocation satisfies both the private measurability condition (if we really want to avoid the mediator) and the Bayesian incentive compatibility condition 4.1.1, and an *ex ante* competitive allocation satisfies only the private measurability condition. It is not clear if Forges, Heifetz and Minelli's results still remain to be true when the two conditions are discriminatorily applied as suggested here. Finally, we repeat our position that a mixed-choice approach avoids the question of explaining deterministic choice. □

Einy, Moreno and Shitovitz (2001b) looked at two notions of the bargaining set of a Bayesian pure exchange economy with a nonatomic measure space of consumers, and established an equivalence result and a nonequivalence result, respectively, with respect to Radner's *ex ante* competitive allocations of the state-contingent market.

Part IV

ANOTHER VIEWPOINT

Chapter 13

Self-Selection in Anonymous Environments

So far we have looked at analyses of situations where several players form a coalition within which to communicate each other and coordinate their strategy choice. Each member of a coalition knows the membership of his coalition, so he knows whom to deal with. In this chapter, however, we will briefly present two models that are based on another view on coalition formation; specifically they describe situations in which coalitional membership is anonymous.

13.1 Mechanism Design

Let $(A, \mathcal{A}, \nu)$ be a probability space of players. Demange and Guesnerie (2001) postulated that there exists a finite state space Ω. Each player a's type is described by a member $\bar{\omega}(a)$ of Ω; a state is synonymously called a type here. The profile of the player types, $\bar{\omega} : A \to \Omega$, is not known, but its distribution $\bar{\pi} := \nu \circ (\bar{\omega})^{-1}$ is public information as the *ex ante* objective probability on the types, and is identified with the grand coalition A since the players are anonymous. Let $\bar{\omega}|_S : S \to \Omega$ be the restriction of $\bar{\omega}$ to coalition $S \in \mathcal{A}$. Associated with each coalition S is the measure on Ω, $\bar{\pi}^S := \nu \circ (\bar{\omega}|_S)^{-1}$, representing the distribution of the members' types. Notice that

$$\bar{\pi}^S(\Omega) = \nu \circ (\bar{\omega}|_S)^{-1}(\Omega) = \nu(S),$$

so the number $\bar{\pi}^S(\Omega)$ is called the size of coalition S.

Demange and Guesnerie studied mechanism design problems for environments in which coalitional membership (precise identity of coalition S)

is not known, but the size (a number $s \in [0,1]$) or type-distribution (a measure π on Ω for which $\pi(\Omega) \leq 1$) are known. Define, therefore, the *space of type distributions*,

$$\Pi := \{\pi : 2^\Omega \to \mathbf{R}_+ \mid \pi \text{ is additive}, \pi(\emptyset) = 0, \pi(\Omega) \leq 1\}.$$

The *interim* stage is defined as the period in which each player a knows his type $\bar{\omega}(a)$ as his private information, as well as the public information $\bar{\pi}$ about the grand coalition.

Denote by C the outcome space. The preference relation of the players of type ω is represented by a von Neumann-Morgenstern utility function, $u(\omega, \cdot) : C \to \mathbf{R}$.

A *plan* associates with each type ω an outcome that will be chosen for ω; a plan is a point in C^Ω. Denote by F^π ($\subset C^\Omega$) the set of all feasible plans when type distribution π prevails; the sets F^π, $\pi \in \Pi$, are *a priori* given to the model. Notice that F^π is defined not only for all probabilities (π for which $\pi(\Omega) = 1$) but also for all measures π for which $\pi(\Omega) \leq 1$. For example, if an unidentified coalition's size s is the only public information about it, a mechanism for this coalition needs to specify a plan in F^π for all π for which $\pi(\Omega) = s$.

The model of this section is summarized as a list of exogenous data,

$$((A, \mathcal{A}, \nu), \ \Omega, \ \bar{\omega}(\cdot), \ C, \ \{F^\pi\}_{\pi \in \Pi}, \{u(\omega, \cdot)\}_{\omega \in \Omega}).$$

A *mechanism* is formulated as a function $f : \Omega \times \Pi \to C$. There are several scenarios for how a mechanism works. The following is a typical one, and gives rise to the u^*-belief-based core concepts introduced below (definitions 13.1.4 and 13.1.5): An underwriter of a mechanism (e.g., the government, or an insurer, or a mediator) designs mechanism f for a yet unidentified coalition, and announces it in public. He then receives players' confidential responses about their private types and about their intention in regard to holding membership in the coalition. A coalition is thus formed. The underwriter forms the distribution of the reported types π of the members. The player who has reported type ω as well as his intention to hold a membership will then receive outcome $f(\omega, \pi)$. The Demange-Guesnerie mechanism can be called an *anonymous* mechanism, in that the outcome for this player is not determined by the exact identities of all respondents or by the exact reported type of each respondent, but merely by the reported-type distribution in addition to his own report.

The grand coalition A has the observable size $\bar{\pi}(\Omega) = 1$, so define $\Pi(1) := \{\pi \in \Pi \mid \pi(\Omega) = 1\}$.

DEFINITION 13.1.1 Mechanism f is *feasible* for A, if

$$\forall \ \pi \in \Pi(1) : f(\cdot, \pi) \in F^\pi.$$

It is called *Bayesian incentive-compatible* for A, if

$$\forall \, \omega, \omega' \in \Omega : \forall \, \pi \in \Pi(1) : u(\omega, f(\omega, \pi)) \geq u(\omega, f(\omega', \pi)).$$

Likewise, these two conditions can be defined for any coalition whose size $s \in [0, 1]$ is public information: Define $\Pi(s) := \{\pi \in \Pi \mid \pi(\Omega) = s\}$. Feasibility and Bayesian incentive compatibility of a mechanism for coalition of size s are defined exactly as in definition 13.1.1, where $\Pi(s)$ substitutes $\Pi(1)$. A feasible mechanism f is called *universally Bayesian incentive-compatible*, if it is Bayesian incentive-compatible for all sizes $s \in [0, 1]$.

REMARK 13.1.2 An alternative definition of feasibility and Bayesian incentive compatibility: A coalition of size 1 is necessarily the grand coalition A (modulo the null sets of players), and the grand coalition's type-distribution $\bar{\pi}$ is public information. So, a mechanism f is *feasible* for a coalition of size 1, if $f(\cdot, \bar{\pi})$ is a member of $F^{\bar{\pi}}$. It is called *Bayesian incentive-compatible* for a coalition of size 1, if

$$\forall \, \omega, \omega' \in \Omega : u(\omega, f(\omega, \bar{\pi})) \geq u(\omega, f(\omega', \bar{\pi})).$$

For any coalition, the weight of its type-ω members cannot exceed $\bar{\pi}(\omega)$. So for coalition S of size $s \leq 1$, the set of all possible type-distributions are

$$\bar{\Pi}(s) := \{\pi \in \Pi(s) \mid \forall \, \omega \in \Omega : \pi(\omega) \leq \bar{\pi}(\omega)\}.$$

A mechanism f is *feasible* for S, if

$$\forall \, \pi \in \bar{\Pi}(s) : f(\cdot, \pi) \in F^{\pi}.$$

It is called *Bayesian incentive-compatible* for S, if

$$\forall \, \omega, \omega' \in \Omega : \forall \, \pi \in \bar{\Pi}(s) : u(\omega, f(\omega, \pi)) \geq u(\omega, f(\omega', \pi)).$$

The above definition and also definition 13.1.1 reflect the scenario that the only public information about a coalition is its size. Actually, there are several variants of these conditions, each reflecting the nature of public information available in a specific context; see, e.g., definitions 13.1.3-13.1.5. □

Demange and Guesnerie proposed several coalitional stability concepts for a feasible, Bayesian incentive-compatible mechanism f for A, each reflecting a specific content of public information, and defined the associated core concepts. The following are a list of possible public information about an unidentified coalition:

- the size $s \in [0, 1]$ of the respondents who express participation in the coalition for which a mechanism is announced;
- the support $\operatorname{supp} \pi \subset \Omega$ of the reported type-distribution;
- the reported type-distribution $\pi \in \Pi$;
- each type ω's reservation level of utility for the status quo, which may be exogenous or endogenous.

One stability concept assumes the situation in which each member of a blocking coalition knows the type-distribution π in his coalition, although he may not know the precise membership of the coalition.

DEFINITION 13.1.3 Let $f : \Omega \times \Pi \to C$ be a standing mechanism for the grand coalition A. Assume that each member of a coalition knows the type-distribution π in the coalition. The coalition *statistically blocks* the mechanism f, if there is a mechanism g that is feasible for π,

$$g(\cdot, \pi) \in F^\pi,$$

and is Bayesian incentive-compatible for π,

$$\forall\, \omega, \omega' \in \Omega : u(\omega, g(\omega, \pi)) \geq u(\omega, g(\omega', \pi)),$$

such that it improves upon f on the support of π,

$$\forall\, \omega \in \operatorname{supp} \pi : u(\omega, g(\omega, \pi)) > u(\omega, f(\omega, \bar{\pi})).$$

The *statistical core* is the set of feasible, Bayesian incentive-compatible mechanisms for A that are not statistically blocked.

Another stability concept, the original conceptual contribution by Demange and Guesnerie (2001), and independently by Hara (2002) (see section 13.2), addresses the situation in which each member of a blocking coalition S knows his reservation utility level, in addition to the type-distribution in S, $\bar{\pi}^S$. The blocking mechanism provides a self-selection criterion to reveal his private information. Here, the blocking coalitions are postulated to be of the following specific form: Let E be any subset of the type space Ω. The E-*full coalition* consists of all players whose types are in E, that is, the coalition $S_E := \bar{\omega}^{-1}(E)$. The exact membership of S_E is not known, since the function $\bar{\omega} : A \to \Omega$ is not known. The type-distribution of the E-full coalition is, however, public information as $\bar{\pi}^{S_E}$, since $\bar{\pi}$ is postulated to be public information. Indeed, $\bar{\pi}^{S_E}(\cdot) = \bar{\pi}(\cdot)$ on the subsets of E. A player of type ω has the reservation level $u^*(\omega)$, in that he will not join a defecting coalition unless his new utility level is made greater than $u^*(\omega)$. In the following blocking criterion, the first set of inequalities says that the players in the blocking coalition S_E improve upon the standing outcomes. The second set of weak inequalities is the self-selection criterion: nobody outside coalition S_E has the incentive to join the blocking coalition. By announcing this mechanism g, the underwriter can form exactly the blocking coalition S_E.

DEFINITION 13.1.4 (Demange and Guesnerie, 2001) Let $f : \Omega \times \Pi \to C$ be a standing mechanism for the grand coalition A. Assume that each player of type ω has his reservation level $u^*(\omega)$. Let $E \subset \Omega$. The E-full coalition u^*-*beliefs blocks* the mechanism f, if there exists a feasible, Bayesian incentive-compatible mechanism g for the type-distribution $\bar{\pi}^{S_E}$ such that

$$\forall\, \omega \in E \quad : \quad u(\omega, g(\omega, \bar{\pi}^{S_E})) > u(\omega, f(\omega, \bar{\pi})), \text{ and}$$
$$\forall\, \omega \in \Omega \setminus E \quad : \quad u(\omega, g(\omega, \bar{\pi}^{S_E})) \leq u^*(\omega).$$

The u^*-*beliefs-based core* is the set of feasible, Bayesian incentive-compatible mechanisms for A that are not u^*-beliefs blocked.

The next blocking concept, and hence the core concept, determines the reservation levels endogenously.

DEFINITION 13.1.5 (Demange and Guesnerie, 2001) Let $f : \Omega \times \Pi \to C$ be a feasible, Bayesian incentive-compatible mechanism for the grand coalition A. Let $E \subset \Omega$. The E-full coalition *status quo blocks* mechanism f, if it u^*-beliefs blocks f, where the reservation levels u^* are given by $u^*(\omega) := u(\omega, f(\omega, \bar{\pi}))$. The *status quo core*indexstatus quo core is the set of feasible, Bayesian incentive-compatible mechanisms for A that are not status quo blocked.

13.2 Pure Exchange Economy

Hara (2002) recently proposed a new core concept for the static pure exchange economy $\mathcal{E}_{PE} := \{\mathbf{R}_+^l, u(a, \cdot), e(a)\}_{a \in A}$ with a nonatomic probability space of consumers $(A, \mathcal{A}, \nu)$; here the type-profile space is a singleton, so notation for a type will be suppressed, and consumer a's preference relation is represented by a utility function of his consumption, $u(a, \cdot) : \mathbf{R}_+^l \to \mathbf{R}$. Hara motivates his new core concept with the imaginary environment in which consumers have gathered in a marketplace, being aware of the statistical distribution of the others' characteristics, so each consumer knows that somewhere in the marketplace there is another consumer he can engage with in a mutually beneficial exchange but he cannot locate such a trading partner. In this environment, he can perhaps post a notice for the entire crowd of consumers to solicit such and such a unit of good A in exchange for such and such a unit of another good B. A coalition can then be formed with whoever comes forward to enter into this trade. (This scenario is also applicable to Demange and Guesnerie's (2001) model in section 13.1.)

For each coalition S, denote by F^S the set of all attainable allocations,

$$F^S := \left\{ f : S \to \mathbf{R}_+^l \;\middle|\; \begin{array}{l} f \text{ is } \nu\text{-integrable,} \\ \int_S f d\nu = \int_S e d\nu \end{array} \right\}.$$

Recall that the net trades of a competitive allocation in $\mathcal{E}_{PE}$ are envy-free. In the following definition 13.2.1, the term $e(a) + (f(b) - e(b))$ is the consumption bundle that consumer a would get if he were to have the same net trade as consumer b. Envy-freeness means that consumer a does not envy the net trades of the other consumers.

DEFINITION 13.2.1 Allocation $f \in F^S$ is called *envy-free*, if there exists $S' \in \mathcal{A}$, $S' \subset S$ and $\nu(S') = \nu(S)$, such that

$$(\forall a, b \in S' : e(a) + (f(b) - e(b)) \in \mathbf{R}_+^l) :$$
$$u(a, f(a)) \geq u(a, e(a) + (f(b) - e(b))).$$

Let $f \in F^A$ be a standing allocation in the grand coalition. Each consumer a's status quo reservation level is then given as $u^*(a) := u(a, f(a))$. The following are analogous to Demange and Guesnerie's (2001) status quo blocking concept and status quo core concept.

DEFINITION 13.2.2 (Hara, 2002) Let $f \in F^A$ be a standing envy-free allocation in the grand coalition A. Coalition S with a positive measure ($S \in \mathcal{A}$, $\nu(S) > 0$) *blocks* allocation f, if there exists an envy-free allocation $g \in F^S$, such that coalition S improves upon f via g, that is, there exists $S' \in \mathcal{A}$, $S' \subset S$ and $\nu(S') > 0$, for which

$$u(a, g(a)) \geq u(a, f(a)), \; \nu\text{-a.e. in } S,$$
$$u(a, g(a)) > u(a, f(a)), \; \nu\text{-a.e. in } S',$$

and such that allocation g satisfies the self-selection criterion *vis-à-vis* f in that no set of outsiders to S with a positive measure want to pretend that they were members of S, that is,

$$\exists S' \in \mathcal{A} : S' \subset S, \; \nu(S') = \nu(S),$$
$$\exists T' \in \mathcal{A} : T' \subset A \setminus S, \; \nu(T') = \nu(A \setminus S),$$
$$(\forall a \in T' : \forall b \in S' : e(a) + (g(b) - e(b)) \in \mathbf{R}_+^l) :$$
$$u(a, e(a) + (g(b) - e(b))) \leq u^*(a) := u(a, f(a)).$$

The *anonymous core* is the set of all envy-free allocations $f \in F^A$ that are not blocked.

Hara's main results are an equivalence theorem between the anonymous core and the set of competitive allocations, and also a generic limit theorem for anonymous core for the replica finite economies.

Bibliography

Allen, B. (1999): "On the existence of core allocations in a large economy with incentive-compatibility constraints," *Fields Institute Communications* **23**, American Mathematical Society, 139-152.

Bahçeci, S. (2003): "The incentive compatible coarse core when information is almost complete," *Journal of Mathematical Economics* **39**, 127-134.

Demange, G., and R. Guesnerie (2001): "On coalitional stability of anonymous interim mechanisms," *Economic Theory* **18**, 367-389.

Einy, E., D. Moreno, and B. Shitovitz (2000a): "On the core of an economy with differential information," *Journal of Economic Theory* **94**, 262-270.

Einy, E., D. Moreno, and B. Shitovitz (2000b): "Rational expectations equilibria and the ex-post core of an economy with asymmetric information," *Journal of Mathematical Economics* **34**, 527-535.

Einy, E., D. Moreno, and B. Shitovitz (2001a): "Competitive and core allocations in large economies with differential information," *Economic Theory* **18**, 321-332.

Einy, E., D. Moreno, and B. Shitovitz (2001b): "The bargaining set of a large economy with differential information," *Economic Theory* **18**, 473-484.

Forges, F., A. Heifetz, and E. Minelli (2001): "Incentive compatible core and competitive equilibria in differential information economies," *Economic Theory* **18**, 349-365.

Forges, F., J.-F. Mertens, and Rajiv Vohra (2002): "The ex ante incentive compatible core in the absence of wealth effects," *Econometrica* **70**, 1865-1892.

Forges, F., and E. Minelli (2001): "A note on the incentive compatible core, "*Journal of Economic Theory* **98**, 179-188.

Forges, F., E. Minelli, and Rajiv Vohra (2002): "Incentives and the core of an exchange economy: a survey, "*Journal of Mathematical Economics* **38**, 1-41.

Glycopantis, D., A. Muir, and N. C. Yannelis (2001): "Am extensive form interpretation of the private core," *Economic Theory* **18**, 293-319.

Hahn, G., and N. C. Yannelis (1997): "Efficiency and incentive compatibility in differential information economies," *Economic theory* **10**, 383-411.

Hahn, G., and N. C. Yannelis (2001): "Coalitional Bayesian Nash implementation in differential information economies," *Economic Theory* **18**, 485-509.

Hara, C. (2002): "The anonymous core of an exchange economy," *Journal of Mathematical Economics* **38**, 91-116.

Hervé-Beloso, C., E. Moreno-Garcia, and N. C. Yannelis (2005): "An equivalence theorem for a differential information economy," *Journal of Mathematical Economics* **41**, 844-856.

Holmström, B., and R. B. Myerson (1983): "Efficient and durable decision rules with incomplete information," *Econometrica* **41**, 1799-1819.

Ichiishi T. (1995): "Cooperative processing of information," in: T. Maruyama and W. Takahashi (eds.), *Nonlinear and Convex Analysis in Economic Theory*, Lecture Notes in Economics and Mathematical Systems, No. 419, pp. 101-117. Heidelberg/Berlin: Springer-Verlag.

Ichiishi T., and A. Idzik (1996): "Bayesian cooperative choice of strategies," *International Journal of Game Theory* **25**, 455-473.

Ichiishi, T., A. Idzik, and J. Zhao (1994): "Cooperative processing of information via choice at an information set," *International Journal of Game Theory* **23**, 145-165.

Ichiishi, T., and S. Koray (2000): "Job matching: A multi-principal, multi-agent model," *Advances in Mathematical Economics* **2**, 41-66.

Ichiishi, T., and R. Radner (1999): "A profit-center game with incomplete information," *Review of Economic Design* **4**, 307-343.

Ichiishi, T., and M. Sertel (1998): "Cooperative *interim* contract and recontract: Chandler's M-form firm," *Economic Theory* **11**, 523-543.

Ichiishi, T., and A. Yamazaki (2002): "Preliminary results for cooperative extensions of the Bayesian game," Discussion Paper No. 2001-9, Graduate School of Economics, Hitotsubashi University, July. Revised: April 2003.

Ichiishi, T., and A. Yamazaki (2004): "*Interim* core concepts for a Bayesian pure exchange economy," *Journal of Mathematical Economics* **40**, 347-370.

Koutsougeras, L. C., and N. C. Yannelis (1993): "Incentive compatibility and information superiority of the core of an economy with differential information," *Economic Theory* **3**, 195-216.

Koutsougeras, L. C., and N. C. Yannelis (1999): "Bounded rational learning in differential information economies: Core and value," *Journal of Mathemtical Economics* **31**, 373-391.

Krasa, S.(1999): "Unimprovable allocations in economies with incomplete information," *Journal of Economic Theory* **87**, 144-168.

Krasa, S., and W. Shafer (2001): "Core concepts in economies where information is almost complete," *Economic Theory* **18**, 451-471.

Lee, D., and O. Volij (2002): "The core of economies with asymmetric information: an axiomatic approach," *Journal of Mathematical Economics* **38**, 46-63.

Lefebvre, I. (2001): "An alternative proof of the nonemptiness of the private core," *Economic Theory* **18**, 275-291.

Maus, S. (2003): "Balancedness and the core in economies with asymmetric information," *Economic Theory* **22**, 613-627.

Maus, S. (2004): "Exchange economies with asymmetric information: Competitive equilibrium and core," *Economic Theory* **24**, 395-418.

Page, Jr., F. H. (1997): "Market games with differential information and infinite dimensional commodity spaces: The core," *Economic Theory* **9**, 151-159.

Page, Jr., F. H., and M. H. Wooders (1994): "Asymmetric information, the efficient core, and farsightedly stable trading mechanisms," unpublished paper.

Serfes, K. (2001): "Non-myopic learning in differential information economies: the core," *Economic Theory* **18**, 333-348.

Serrano, R., and Rajiv Vohra (2001): "Some limitations of virtual Bayesian implementation," *Econometrica* **69**, 785-792.

Serrano, R., Rajiv Vohra, and O. Volij (2001): "On the failure of core convergence in economies with asymmetric information," *Econometrica* **69**, 1685-1696.

Vohra, Rajiv (1999): "Incomplete information, incentive compatibility, and the core," *Journal of Economic Theory* **86**, 123-147.

Volij, O. (2000): "Communication, credible improvements and the core of an economy with asymmetric information," *International Journal of Game Theory* **29**, 63-79.

Wilson, R. (1978): "Information efficiency, and the core of an economy," *Econometrica* **46**, 807-816.

Yannelis, N. C. (1991): "The core of an economy with differential information," *Economic Theory* **1**, 183-198.

Yazar, J. (2001): "Ex ante contracting with endogenously determined communication plans," *Economic Theory* **18**, 439-450.

Related Works

Abreu, D., and H. Matsushima (1992): "Virtual implementation in iteratively undominated strategies: Complete information," *Econometrica* **60**, 993-1008.

Akerlof, G. (1970): "The market for lemons: Quality uncertainty and the market mechanisms," *Quarterly Journal of Economics* **84**, 488-500.

Aumann, R. J. (1964): "Markets with a continuum of traders," *Econometrica* **32**, 39-50.

Aumann, R. J. (1966): "Existence of competitive equilibria in market with a continuum of traders," *Econometrica* **34**, 1-17.

Aumann, R. J. and B. Peleg (1960): "Von Neumann-Morgenstern solutions to cooperative games without side payments," *Bulletin of the American Mathematical Society* **66**, 173-179.

Chandler, Jr., A. D. (1962): *Strategy and Structure*, Cambridge, MA: MIT Press.

d'Aspremont, C., and L.-A. Gérard-Varet (1979): "Incentives and incomplete information," *Journal of Public Economics* **11**, 22-45.

Debreu, G., and H. Scarf (1963): "A limit theorem on the core of an economy," *International Economic Review* **4**, 235-246.

Harsanyi, J. C. (1967/1968): "Games with incomplete information played by 'Bayesian' players," *Management Science: Theory* **14**, 159-182 (Part I), 320-334 (Part II), 486-502 (Part III).

Hildenbrand, W. (1968): "On the core of an economy with a measure space of economic agents," *Review of Economic Studies* **35**, 443-452.

Hildenbrand, W. (1970): "Existence of equilibria for economies with production and a measure space of consumers," *Econometrica* **38**, 608-623.

Ichiishi, T. (1981): "A social coalitional equilibrium existence lemma," *Econometrica* **49**, 369-377.

Ichiishi, T. (1993a): *The Cooperative Nature of the Firm*, Cambridge, U.K.: Cambridge University Press.

Ichiishi, T. (1993b): "The cooperative nature of the firm: Narrative," *Managerial and Decision Economics* **14** (1993), 383-407. In: Special Issue edited by Koji Okuguchi on Labor-Managed Firms Under Imperfect Competition (and Related Problems).

Jackson, M. (1991): "Bayesian implementation," *Econometrica*, **59**, 461-477.

Klein, B., R. Crawford, and A. A. Alchian (1978): "Vertical integration, appropriable rents, and the competitive contracting process," *Journal of Law and Economics* **21**: 297-326.

Koray, S., and M. R. Sertel (1992): "The welfarisitc characterization of two-person revelation equilibrium under imputational government," *Social Choice and Welfare*, **9**, 49-56.

Kreps, D., and R. Wilson (1982): "Sequential equilibria," *Econometrica* **50**, 863-894.

Mas-Colell, A. (1977): "Indivisible commodities and general equilibrium theory," *Journal of Economic Theory* **16**, 443-456.

Myerson, R. B. (1984): "Cooperative games with incomplete information," *International Journal of Game Theory* **13**, 69-96.

Radner, R. (1968): "Competitive equilibrium under uncertainty," *Econometrica* **36**, 31-58.

Radner, R. (1979): "Rational expectations equilibrium generic existence and the information revealed by prices," *Econometrica* **47**, 655-678.

Radner, R. (1992): "Transfer payments and the core of a profit-center game," in: P. Dasgupta, *et al.* (eds.), *Economic Analysis of Markets and Games (Essays in Honor of Frank Hahn)*, pp. 316-339. Cambridge, MA: MIT Press.

Rosenmüller, J. (1992): "Fee games: (N)TU-games with incomplete information," in: R. Selten (ed.) *Rational Interaction: Essays in Honor of John C. Harsanyi*, pp.53-81. Berlin: Springer-Verlag.

Scarf, H. (1973): *The Computation of Economic Equilibria*, New Haven, CT: Yale Univ. Press.

Scarf, H. (1986): "Notes on the core of a productive economy," in W. Hildenbrand, *et al.* (eds.), *Contributions to Mathematical Economics (in Honor of Gerard Debreu)*, pp. 401-429. Amsterdam/New York: North-Holland.

Selten, R. (1975): "Reexamination of the perfectness concept for equilibrium points in extensive games," *International Journal of Game Theory* **4**, 25-55.

Vind, K. (1972): "A third remark on the core of an atomless economy," *Econometrica* **40**, 585-586.

Walkup, D. W., and R. J.-B. Wets (1969):"Lifting projections of convex polyhedra," *Pacific Journal of Mathematics* **28**, 465-475.

Williamson, O. E. (1975): *Markets and Hierarchies: Analysis and Antitrust Implications*, Free Press, New York, NY.

Yamazaki, A. (1978): "An equilibrium existence theorem without convexity assumptions," *Econometrica* **46**, 541-555.

Yamazaki, A. (1981): "Diversified consumption characteristics and conditionally dispersed endowment distribution: Regularizing effect and existence of equilibria," *Econometrica* **49**, 639-654.

Glossary

$\{\mathcal{A}^j\}_{j\in S}$: communication system for coalition S

$\mathcal{D} := (\mathcal{E}_{cp}, p)$: profit-center game with incomplete information, where $\mathcal{E}_{cp}$ is a Bayesian coalition production economy (see below), and p is a price vector for the marketed commodities.

$\mathcal{E}_{cp} := (\{X^j, T^j, u^j, e^j, \{\pi^j(\cdot \mid t^j)\}_{t^j \in T^j}\}_{j \in N}, \{Y^S\}_{S \in \mathcal{N}})$: Bayesian coalition production economy with l commodities, where:

$\{X^j, T^j, u^j, e^j, \{\pi^j(\cdot \mid t^j)\}_{t^j}\}_j =: \mathcal{E}_{pe}$, the consumption sector, i.e., Bayesian pure exchange economy;

$\{Y^S\}_{S \in \mathcal{N}}$: the production sector. Correspondence $Y^S : T \to \mathbf{R}^l$ associates to each type profile t a production set $Y^S(t)$ ($\subset \mathbf{R}^l$) for coalition S.

$\mathcal{E}_{ks}(\pi) := (\Omega, \pi, \{\mathbf{R}_+^l, u^j, e^j\}_{j \in N})$: K-S pure exchange economy with l commodities, where:

N: a finite set of consumers;
$\mathbf{R}_+^l$: consumption set of consumer j;
Ω: finite state space;
$u^j : \mathbf{R}_+^l \times \Omega \to \mathbf{R}$, state dependent von Neumann-Morgenstern utility function of consumer j;
$e^j : \Omega \to \mathbf{R}_+^l$, consumer j's initial endowment vector, which depends only upon ω;
π: objective *ex ante* probability on $\Omega \times \prod_{j \in N} \Phi^j$, where Φ^j ($= \Omega$) is the consumer j's signal space.

$\mathcal{E}_{pe} := \{X^j, T^j, u^j, e^j, \{\pi^j(\cdot \mid t^j)\}_{t^j \in T^j}\}_{j \in N}$: Bayesian pure exchange economy with l commodities, where:

N: a finite set of consumers;
X^j ($\subset \mathbf{R}_+^l$): consumption set of consumer j;
T^j: finite type set of consumer j;
$u^j : \mathbf{R}_+^l \times T \to \mathbf{R}$, type-profile dependent von Neumann-Morgenstern utility function of consumer j;
$e^j : T^j \to \mathbf{R}_+^l$, consumer j's initial endowment vector, which depends only upon t^j;
$\pi^j(\cdot \mid t^j)$: consumer j's conditional probability on $T^{N \setminus \{j\}}$ given t^j, objective or subjective.

$\mathcal{E}_{PE} := ((A, \mathcal{A}, \nu), \{\mathbf{R}_+^l, T^{j(a)}, u(a, \cdot, \cdot), e(a, \cdot), \pi^a\}_{a \in A})$: Bayesian pure exchange economy with a nonatomic measure space of consumers $(A, \mathcal{A}, \nu)$.

$F'^S : X \to X^S$, $\bar{x} \mapsto \{x^S \in F^S(\bar{x}) \mid \forall j \in S : x^j \text{ is } \mathcal{T}^j\text{-measurable.}\}$, private-measurable and feasible strategy set correspondence for coaltion S in a Bayesian society $\mathcal{S}$ (defined for the private information case).
$\hat{F}^S : X \to X^S$, $\bar{x} \mapsto \{x^S \in F'^S(\bar{x}) \mid \forall j \in S : x^j \text{ is Bayesian incentive-compatible.}\}$, private-measurable, feasible and Bayesian incentive-compatible strategy set correspondence for coaltion S in a Bayesian society $\mathcal{S}$ (defined for the private information case).

$\mathbf{R}^N$: #N-dimensional Euclidean space, where N is the index set for the coordinates. For $x, y \in \mathbf{R}^N$, and $S \subset N$,

$x \geq y \iff \forall j \in N : x_j \geq y_j$;
$x > y \iff [x \geq y \text{ and } x \neq y]$;
$x \gg y \iff \forall j \in N : x_j > y_j$.
χ_S ($\in \mathbf{R}^N$): the characteristic vector of S, defined by $(\chi_S)_j = 1$ if $j \in S$, $(\chi_S)_j = 0$ if $j \in N \setminus S$.

$\mathcal{S} := \left(\{C^j, T^j, u^j, \{\pi^j(\cdot \mid t^j)\}_{t^j \in T^j}\}_{j \in N}, \{\mathbf{C}_0^S, T(S), F^S\}_{S \in \mathcal{N}}\right)$: Bayesian society with *interim* probabilities, where:

N: a finite set of players;

$\mathcal{N} := 2^N \setminus \{\emptyset\}$, the family of nonempty coalitions;
C^j: a choice set of player j;
$C^S := \prod_{j \in S} C^j$;
$C := C^N$, the set of choice bundles;
T^j: a finite set of types of player j;
$T^S := \prod_{j \in S} T^j$, the set of type profiles for coalition S;
$T := T^N$;
$u^j : C \times T \to \mathbf{R}$, a type-profile dependent von Neumann-Morgenstern utility function of player j;
$\pi^j(\cdot \mid t^j)$: player j's conditional probability on the others' type profiles $T^{N \setminus \{j\}}$, given private information $t^j \in T^j$, objective or subjective;
$\mathbf{C}_0^S : T \to C^S$, a feasible-choice correspondence of coalition S;
$T(S)$: the domain of strategies of coalition S;
$X^j(S) := \{x^j : T(S) \to C^j\}$, the set of all logically conceivable strategies of player j as a member of S;
$X^S := \prod_{j \in S} X^j(S)$, the set of all logically conceivable strategy bundles of coalition S, in particular,
$X^{\{j\}} := X^j(\{j\})$, the set of all logically conceivable strategies of singleton $\{j\}$;
$X := X^N$;
$F^S : X \to X^S$: a feasible-strategy correspondence of coalition S (a family of $\mathcal{T}^S$-measurable selections of $\mathbf{C}_0^S|_{T(S)}$).

supp $\pi^j(\cdot \mid t^j) := \{t^{N \setminus \{j\}} \in T^{N \setminus \{j\}} \mid \pi^j(t^{N \setminus \{j\}} \mid t^j) > 0\}$, the support of *interim* probability $\pi^j(\cdot \mid t^j)$. Sometimes supp $\pi^j(\cdot \mid t^j)$ may be considered a subset of $\{t^j\} \times T^{N \setminus \{j\}}$; no confusion arises.

$T(\pi^j) := \bigcup_{t^j \in T^j} \{t^j\} \times$ supp $\pi^j(\cdot \mid t^j)$, the set of type profiles with a positive probability,
$T(S)$: the domain of strategy bundles of coalition S,
$\text{proj}_j T(S)$: projection of $T(S)$ to T^j, for $j \in S$.

$\mathcal{T}^S$: the algebra on T generated by partition, $\{\{t^S\} \times T^{N \setminus \{S\}} \mid t^S \in T^S\}$,
$\mathcal{T}^j := \mathcal{T}^{\{j\}}$: player j's private information structure,
$\mathcal{T}^j(\pi^j) := \mathcal{T}^j \bigvee \{\emptyset, T(\pi^j), T \setminus T(\pi^j), T\}$.

Note: For the K-S pure exchange economy, T^j is defined differently; see section 7.2.

Index

A

Abreu, Dilip, 30
agent, 91
Alchian, Armen A., 119
Allen, Beth, 205
anonymous core, 225
asset specificity, 119
Aumann, Robert J., 3, 10, 98, 99, 195, 212

B

balanced family, 99
balanced non-side-payment game, 99
balancing coefficients, 99
bargaining set, 215
Bayesian game
— , 8
— , cooperative extensions of, 4
Bayesian coalition production economy, 14
Bayesian incentive-compatible
— (for the mediator-based approach), 37, 41
— (for the private information case), 27, 29
— (for the private information case), strictly, 30
— (for the two-*interim*-period case), 151
— in Krasa and Shafer's sense, 73
— in a large Bayesian pure exchange economy, 198
Bayesian incentive-compatible coarse core (for the mediator-based approach), 49
Bayesian incentive-compatible coarse core (for the private information case), 49, 168
Bayesian incentive-compatible coarse efficient, 59
Bayesian incentive-compatible coarse strong equilibrium (for the mediator-based approach), 49
Bayesian incentive-compatible coarse strong equilibrium (for the private information case), 48
Bayesian incentive-compatible core, *ex ante*, 53
Bayesian incentive-compatible *interim* efficient, 60, 62
Bayesian incentive-compatible strategy bundle pair, 146
Bayesian incentive-compatible strong equilibrium, 112
Bayesian pure exchange economy, 13
— , large, 196
Bayesian society, 11
— (no-externalities case), 94

C

Chandler, Jr., Alfred D., 14, 118
coalition structure, 55, 62
coarse core
— , 47
— (for the mediator-based approach), Bayesian incentive-compatible, 49
— (for the private information case), Bayesian incentive-compatible, 49

coarse efficient, 58
coarse strong equilibrium
—, 46
— (for the mediator-based approach), Bayesian incentive-compatible, 49
— (for the private information case), Bayesian incentive-compatible, 48
communication plan, 40
communication system
—, 12
—, full, 13
—, null, 13
— for a large Bayesian pure exchange economy, 67
— for a large Bayesian pure exchange economy, full, 68
complete information, 9
— in Krasa and Shafer's sense, 72
commodity
—, marketed, 15, 119
—, nonmarketed, 15, 119
complementary supplier - customer relationship, 129
conditional expected utility function, 46
constrained market equilibrium, 209
contract, 52
convex technology, 125
cooperative extensions of the Bayesian game, 4
core, 99
—, anonymous, 225
—, coarse, 47
—, EC- (endogenous communication plan), 148
—, *ex ante* Bayesian incentive-compatible, 53
—, fine, 50
—, generalized, 100
—, *interim* Bayesian incentive-compatible, 51, 172
—, statistical, 222
—, strict, 81

—, u^*-beliefs-based, 223
— (for the mediator-based approach), Bayesian incentive-compatible coarse, 49
— (for the private information case), Bayesian incentive-compatible coarse, 49
— of a K-S pure exchange economy, Bayesian incentive-compatible, 74
— of a K-S pure exchange economy, complete information, 72
— of a K-S pure exchange economy, private information, 73
— of a large Bayesian pure exchange economy, *ex ante* Bayesian incentive compatible, 203, 205
— of a large Bayesian pure exchange economy, *ex ante* private information, 202
— of a large Bayesian pure exchange economy, *interim* Bayesian incentive-compatible, 198
Crawford, Robert G., 119

D

d'Aspremont, Claude, 27
Debreu, Gerard, 207
Demange, Gabrielle, 223
distributive production set, 127
durable strategy bundle, 147

E

EC-core (endogenous communication plan core), 148
economy
—, Bayesian coalition production, 14
—, Bayesian pure exchange, 13
—, K-S pure exchange, 72
—, large Bayesian pure exchange, 196
—, replica, 207
efficient

—, Bayesian incentive-compatible coarse, 59
—, Bayesian incentive-compatible *interim*, 60, 62
—, coarse, 58
—, *ex ante*, 58
—, fine, 59, 60
—, *interim* private, 60
—, *ex post*, 59
E-full coalition, 222
Einy, Ezra, 53, 69, 196, 197, 203, 212, 215
envy-free allocation, 224
equilibrium
  —, Bayesian incentive-compatible strong, 112
  —, constrained market, 209
  —, Nash, 101
  —, perfect, 144
  —, rational expectations, 54, 110
  —, sequential, 144
  —, social coalitional, 102
  —, strong, 102
  — of the state-contingent claim market, *ex ante* competitive, 202
ex ante, 8
ex ante Bayesian incentive-compatible core, 53
  — of a large Bayesian pure exchange economy, 203, 205
ex ante efficient, 58
ex ante competitive equilibrium of the state-contingent claim market, 202
ex ante private information core of a large Bayesian pure exchange economy, 202
ex ante solution, 45
ex post, 8
ex post core of a large Bayesian pure exchange economy, 68
ex post efficient, 59

F

fine core, 50

— of a large Bayesian pure exchange economy, 68
fine efficient, 59, 60
fine strong equilibrium, 50
firm in multidivisional form, 14
Forges, Françoise, 23, 190, 192, 214
full communication system, 13
fully pooled information case, 23

G

game
  —, balanced non-side-payment, 99
  —, Bayesian, 8
  —, non-side-payment, 10, 98
  —, non-transferable utility, 10, 98
  —, NTU, 10, 98
  —, profit center, 15, 118
  — in characteristic function form without side payments, 98
  — in normal form, 101
generalized core, 100
generic property of a K-S pure exchange economy, 75
Gérard-Varet, Louis-André, 27
Guesnerie, Roger, 223

H

Hahn, Guangsug, 31, 60, 169, 172
Hara, Chiaki, 224
Harsanyi, John C., 3, 8
headquarters' insurability, postulate of, 122
Heifetz, Aviad, 214
Holmström, Bengt, 59, 138

I

Ichiishi, Tatsuro, 11, 15, 29, 51, 52, 53, 54, 61, 94, 102, 104, 105, 115, 117, 125, 128, 129, 130, 134, 155, 168, 169, 177, 188
Idzik, Adam, 11, 29, 53, 94, 115, 117, 188

information
— , complete, 9
— , non-exclusive, 170
— , private, 8
— in Krasa and Shafer's sense, complete, 72
— in Krasa and Shafer's sense, incomplete, 71
— in Krasa and Shafer's sense, private, 72
information-revelation process, postulate of, 111
information structure, private, 8, 16
— in Krasa and Shafer's sense, 73
in mediis, 8
interim, 8
interim Bayesian incentive-compatible core, 51, 172
— of a large Bayesian pure exchange economy, 198
interim Bayesian incentive-compatible strong equilibrium, 51
interim private efficient, 60
interim solution, 45

J

Jackson, Matthew, 17

K

Klein, Benjamin, 119
Koray, Semih, 54, 150
Krasa, Stefan, 75, 76, 81
Kreps, David M., 144
K-S pure exchange economy, 72

L

large Bayesian pure exchange economy, 196
Lefebvre, Isabelle, 94

M

market for a lemon, 172
marketed commodity, 15, 119
Matsushima, Hitoshi, 30
mechanism, 220
— , Bayesian incentive-compatible, 221
— , feasible, 220, 221
mediator, 37
mediator-based approach, 37
Mertens, Jean-François, 192
M-form firm, 14
Minelli, Enrico, 23, 190, 214
Moreno, Diego, 53, 69, 196, 197, 203, 212, 215
multi-principal, multi-agent relationship, 54
Myerson, Roger B., 29, 59, 138

N

Nash equilibrium, 101
no-externality case, 31, 57
non-exclusive information, 170
nonmarketed commodity, 15, 119
non-side-payment game, 10, 98
— , balanced, 99
non-transferable utility game, 10, 98
NTU game, 10, 98
null communication system, 13

P

parameterized family of societies, 106
— , social coalitional equilibrium of, 106
Peleg, Bezalel, 3, 10, 98, 99
perfect equilibrium, 144
plan, 10
postulate of the headquarters' insurability, 122
postulate of information-revelation process, 111
pretend-but-perform principle, 150

pretension function, 29, 150
private information, 8
private information case, 22
private information structure, 8, 16
— in Krasa and Shafer's sense, 73
private measurability, 22
profit center game, 15, 61, 118

Q

Quinzii, Martine, 155

R

Radner, Roy, 15, 23, 110, 118, 125, 128, 129, 130, 169, 202, 211, 215
rational expectations equilibrium, 54, 110
replica economy, 207
revelation principle, 35
Rosenmüller, Joachim, 57

S

Scarf, Herbert E., 3, 100, 127, 133, 207 149, 155, 169
Selten, Reinhard, 142, 144, 145
sequential equilibrium, 144
Serrano, Roberto, 209
Sertel, Murat R., 51, 52, 61, 134, 150
Shafer, Wayne, 75, 76, 81
Shitovitz, Benyamin, 53, 69, 196, 197, 203, 212, 215
social coalitional equilibrium, 102
society, 102
—, Bayesian, 11
— (no-externalities case), 105
solution
—, *ex ante*, 45
—, *interim*, 45
space of Bayesian societies, 116
statistical core, 222
strict core allocation plan, 81
strictly Bayesian incentive-compatible (for the private information case), 30
strong equilibrium, 102
—, coarse, 46
—, fine, 50
—, *interim* Bayesian incentive-compatible, 51
— (for the mediator-based approach), Bayesian incentive-compatible coarse, 49
— (for the private information case), Bayesian incentive-compatible coarse, 48

T

transfer payment problem, 15, 119

U

u^*-beliefs-based core, 223

V

Vind, Karl, 69
Vohra, Rajiv, 23, 37, 49, 169, 170, 172, 190, 192, 209
Vohra box diagram, 172
Volij, Oscar, 209

W

Walkup, David W., 158
Wets, Roger J.-B., 158
Williamson, Oliver E., 119
Wilson, Robert, 3, 12, 22, 46, 49, 58, 59, 91, 144, 169

Y

Yamazaki, Akira, 168, 177, 205
Yannelis, Nicholas C., 22, 31, 52, 60, 93, 169, 172
Yazar, Jülide, 41, 148

Z

Zhao, Jingang, 115, 117